Name: _____

CEM

Non-Verbal Reasoning & Spatial Reasoning

11+ Practice Papers

Peter Francis

GALORE PARK

AN HACHETTE UK COMPANY

Every effort has been made to trace all copyright holders, but if any have been inadvertently overlooked, the Publishers will be pleased to make the necessary arrangements at the first opportunity.

Although every effort has been made to ensure that website addresses are correct at time of going to press, Galore Park cannot be held responsible for the content of any website mentioned in this book. It is sometimes possible to find a relocated web page by typing in the address of the home page for a website in the URL window of your browser.

Hachette UK's policy is to use papers that are natural, renewable and recyclable products and made from wood grown in well-managed forests and other controlled sources. The logging and manufacturing processes are expected to conform to the environmental regulations of the country of origin.

Orders: please contact Hachette UK Distribution, Hely Hutchinson Centre, Milton Road, Didcot, Oxfordshire, OX11 7HH. Telephone: (44) 01235 827827. Email education@hachette.co.uk. Lines are open from 9 a.m. to 5 p.m., Monday to Friday. You can also order through our website: hoddereducation.com.

Parents, Tutors please call: 020 3122 6405 (Monday to Friday, 9:30 a.m. to 4.30 p.m.).
Email: parentenquiries@galorepark.co.uk

Visit our website at www.galorepark.co.uk for details of other revision guides for Common Entrance, examination papers and Galore Park publications.

ISBN: 978 1 5104 4974 9

© Peter Francis 2019
First published in 2019 by
Hodder & Stoughton Limited
An Hachette UK Company
Carmelite House
50 Victoria Embankment
London EC4Y 0DZ
www.galorepark.co.uk
Impression number   10 9 8 7 6 5 4 3 2
Year   2023 2022

Illustrations by Peter Francis

Typeset in India
Printed in the UK

A catalogue record for this title is available from the British Library.

www.carbonbalancedprint.com
CBP2250

Contents and progress record

Section	Page	Length (no. Qs)	Timing (mins)	Question type	Score	Time
Paper 1 Foundation level Representing a CEM test at an average level of challenge for grammar and independent schools.						
Non-verbal reasoning	8	15	7	Multiple choice	/ 15	:
Spatial reasoning	13	22	12	Multiple choice	/ 22	:
Non-verbal and spatial reasoning	20	12	6	Multiple choice	/ 12	:
				Total	/ 49	:
Paper 2 Standard level Representing a CEM test at a medium level of challenge for grammar and independent schools.						
Spatial reasoning	25	12	7	Multiple choice	/ 12	:
Non-verbal and spatial reasoning	28	13	7	Multiple choice	/ 13	:
Non-verbal reasoning	32	24	11	Multiple choice	/ 24	:
				Total	/ 49	:
Paper 3 Advanced level Representing a CEM test at a high level of challenge for grammar and independent schools.						
Non-verbal and spatial reasoning	39	24	10	Multiple choice	/ 24	:
Non-verbal reasoning	46	12	5	Multiple choice	/ 12	:
Spatial reasoning	51	11	5	Multiple choice	/ 11	:
				Total	/ 47	:

Go to the Galore Park website to download the free PDF answer sheets to use and re-use as many times as you need: galorepark.co.uk/answersheets

How to use this book

Introduction

These practice papers have been written to provide final preparation for your CEM 11+ non-verbal reasoning test. To give you the best chance of success, Galore Park has worked with 11+ tutors, independent schools' teachers, test writers and specialist authors to create these practice papers.

This book includes three model papers. Each paper contains between 47 and 49 multiple-choice questions, covering a variety of skills. The papers increase in difficulty from Paper 1 to Paper 3 and all work to a timing that has been typical of CEM tests in the past. This is because CEM tests can change in difficulty both from year to year and from school to school. Since each paper differs slightly in complexity and skill area, we suggest you complete all three papers to help you fully prepare for the challenges ahead.

So that you experience how the CEM tests work, we have included a few key elements to help you become familiar with what to expect:

- Each test lasts between 20 and 25 minutes (and there are often many questions that you may not complete in the time given).
- The sections (or **parts**) within each paper are short and of unpredictable length.
- Parts generally begin with an untimed introduction, an example question and a practice question to explain each question format.

It is important to read the instructions carefully as you will be asked to record your answers in a variety of ways:

- choosing a multiple-choice option using a separate answer sheet
- choosing a multiple-choice option, writing / recording this on the paper itself.

These different styles are included because the online tests expect you to be able to adapt to different question formats quickly as you move on from one part of a paper to the next. These formats can also change from year to year.

As you mark your answers, you will see references to the Galore Park *11+ Non-Verbal Reasoning Study and Revision Guide*. These references have been included so that you can go straight to some useful revision tips and find extra practice questions for those areas where you would like more help.

Working through the book

The **Contents and progress record** on page 3 helps you to track your scores and timings as you work through the papers.

You may find some of the questions hard, but don't worry – these tests are designed to make you think. Agree with your parents on a good time to take the test and follow the instructions below to prepare for each paper as if you are actually going to sit your 11+ non-verbal reasoning test.

1 Read the instructions on page 7 before you begin each practice paper.
2 Download the answer sheet from www.galorepark.co.uk/answersheets and print it out before you begin.
3 Take the test in a quiet room. Set a timer and record your answers as instructed.
4 Note down how long the test takes you (all questions should be completed even if you run over the time suggested). If possible, complete a whole paper in one session.

5 Mark the paper using the answers at the back of the book.

6 Go through the paper again with a friend or parent, talk about the difficult questions and note which parts of the revision guide you are going to review.

The **Answers** can be cut out so that you can mark your papers easily. Do not look at the answers until you have attempted a whole paper.

When you have finished a complete paper, turn back to the **Contents and progress record** and fill in the boxes. Make sure to write your total number of marks and time taken in the **Score** and **Time** boxes.

If you would like to take further CEM-style papers after completing this book, you will find more papers in the *Pre-test/ 11+ Non-Verbal Reasoning Practice Papers 1* and *2* (see **Continue your learning journey** on page 6).

Test day tips

Take time to prepare yourself on the day before you go for the test. Remember to take sharpened pencils, an eraser and, if you are allowed, water to maintain your concentration levels and a watch to time yourself.

... and don't forget to have breakfast before you go!

Pre-test and the 11+ entrance exams

This title is part of the Galore Park *Pre-test/11+* series and there are three further *Non-Verbal Reasoning Practice Paper* titles (see **Continue your learning journey** on page 6).

This series is designed to help you prepare for pre-tests and 11+ entrance exams if you are applying to independent schools. These exams are often the same as those set by local grammar schools.

Pre-tests and 11+ non-verbal reasoning tests appear in a variety of formats and lengths and it is likely that if you are applying for more than one school, you will encounter more than one style of test. These include:

● Pre-test/11+ entrance exams in different formats from GL, CEM and ISEB
● Pre-test/11+ entrance exams created specifically for particular schools.

As the tests change all the time, it can be difficult to predict the questions, making them harder to revise for. If you are taking more than one style of test, review the books in the **Continue your learning journey** section to see which other titles could be helpful to you.

For parents

For your child to get the maximum benefit from these papers, they should complete them in conditions as close as possible to those they will face in the actual test, as described in the **Working through the book** section on page 4.

Working with your child to follow up the revision work suggested in the answers can improve their performance in areas where they are less confident and boost their chances of success.

Continue your learning journey

When you've completed these *Practice Papers*, you can carry on your learning right up until exam day with the following resources.

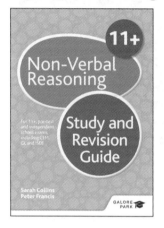

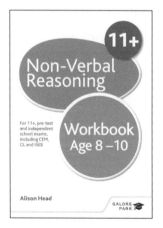
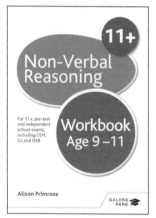
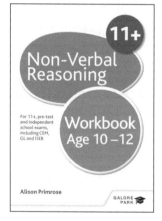

The *Revision Guide* (referenced in the answers to this book) covers basic skills in all areas of non-verbal reasoning, and guidance is provided on how to improve in this subject.

Pre-test/11+ Practice Papers 1 and *2* are designed to provide a complete revision experience across the various test styles you may encounter. Between the two titles there are eighteen tests of varying length, each followed by comprehensive answer explanations.

- *Book 1* begins with four training tests, followed by four short papers and answers designed to develop your confidence, speed and familiarity with taking the actual tests.
- *Book 2* contains a further five model papers and answers to improve your accuracy, speed and ability to deal with variations in question format under pressure.

GL 11+ Non-Verbal Reasoning Practice Papers contains three practice papers designed for preparation for the GL-style tests.

The *Workbooks* will further develop your skills with over 160 questions to practise in each book. To prepare you for the exam, these books include even more of the question variations that you may encounter – the more you do, the better equipped for the exams you will be:

- Age 8–10: Increase your familiarity with variations in the question types.
- Age 9–11: Experiment with further techniques to improve your accuracy.
- Age 10–12: Develop fast response times through consistent practice.

Use Atom Learning to improve familiarity with online tests: the online learning platform adapts to your ability to ensure you are always working on your optimal learning path and the adaptive, mock-testing facility looks and scores in the style of the pre-tests. galorepark.co.uk/atomlearning

Preparing for each paper

Read these instructions before you begin each practice paper.

1   Take the test in a quiet room. Have your timer ready.
2   Check at the beginning of **Part 1** if you will be recording your answers on an **answer sheet**. If a sheet is required, download it from galorepark.co.uk/answersheets and print it out before you begin.
3   The test is made up of three Parts, 1–3. You should complete all three parts of the paper.
4   Parts generally begin with an introduction, followed by example questions and training questions for each question type. These are untimed and so you should read these instructions carefully and then complete the example questions and training questions before beginning the timed questions.
5   Start the timer *after* completing the introduction to each part and before you look at the timed questions.
6   Stop the timer at the end of each part, as instructed.
7   For Parts 1–3:
    a   aim for the time given
    b   complete all questions
    c   note the actual time you have taken at the end of each part.
8   Answer the questions as described in the introduction at the beginning of each part, using a pencil.
9   If you want to change an answer as you work through a part, rub your answer out and rewrite it. You cannot change an answer after you have completed a part.
10 Work as quickly and efficiently as you can. If a question is difficult to answer, come back to it after finishing the other questions in that part.
11 Aim to answer each question before you finish, even if you are not completely sure of the answer.
12 *Do not look at the answers before completing the entire paper.* The instructions in **Working through the book** on page 4 explain how to review your answers.

Always read the instructions on exam papers carefully and make sure you understand exactly what you need to do to answer the questions.

● Paper 1

Test time: 25 minutes

Download and print the answer sheet from galorepark.co.uk/answersheets before you start this paper.

Complete all three parts of this paper according to the timings given at the start of each set of questions. Stop the timer after completing each part and start it again after answering the training questions.

Part 1: Non-verbal reasoning

How to answer these questions

All your answers to this part should be recorded on the answer sheet you have downloaded. Look at the examples and then complete the training questions, which have answers at the end of this introduction. **Do not begin timing yourself until you have finished these pages.**

Only check your answers after completing all of Paper 1. The answers are in a cut-out section at the end of the book. Complete the 'results' boxes at the end of this part when you have added up your score. If you run over the time given, complete the questions and note the time you have taken.

> ### Example question 1
> Look at the first three pictures and decide what they have in common. Then select the option from the five on the right that belongs to the same set.
>
>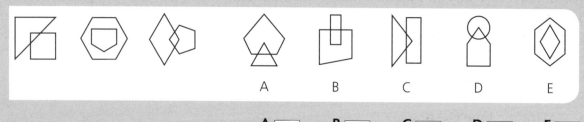
>
> A ☐   B ☐   C ▬   D ☐   E ☐
>
> Answer: **C**   **shape** – each picture is made up of two different shapes; **number** – one of the shapes has one more side than the other.
>
> *Distractors:* **shape** – the shapes are unimportant, as is the overlap; **size** – the size of the shapes is unimportant.

Now answer Training question 1, recording your answer below as shown in the example.

> ### Training question 1
>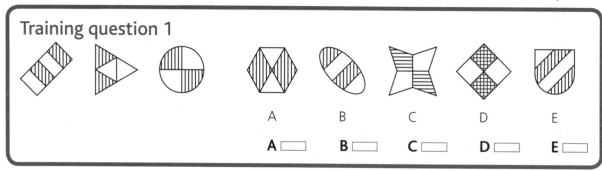
>
> A ☐   B ☐   C ☐   D ☐   E ☐

Example question 2

The boxes on the left show a pattern that is arranged in a sequence. Choose the answer option that completes the sequence when inserted in the blank box.

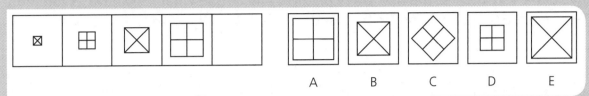

A ⬜ B ⬜ C ⬜ D ⬜ E ⬛

Answer: E   size/proportion – the square grows in size from one box to the next; **shading** – the centre shading alternates between a cross and an 'X'.

Now answer Training question 2, recording your answer below as shown in the example.

Training question 2

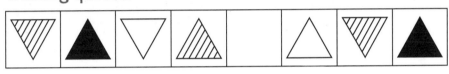

A ⬜ B ⬜ C ⬜ D ⬜ E ⬜

Use the downloaded answer sheet to record your answers to the questions that follow.

You will see that the examples and training questions have already been recorded.

You now have 7 minutes to complete the following 15 questions.

Questions 1–8: Look at the first three pictures and decide what they have in common. Then select the option from the five on the right that belongs to the same set.

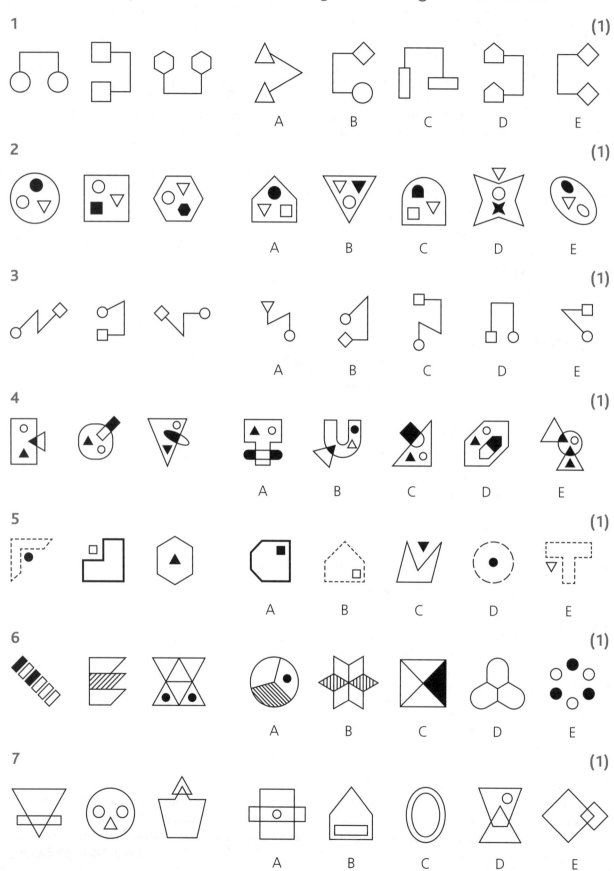

1 (1)

A   B   C   D   E

2 (1)

A   B   C   D   E

3 (1)

A   B   C   D   E

4 (1)

A   B   C   D   E

5 (1)

A   B   C   D   E

6 (1)

A   B   C   D   E

7 (1)

A   B   C   D   E

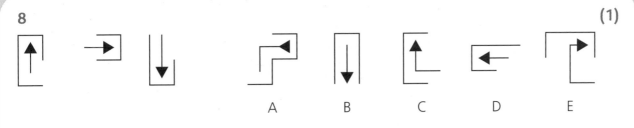

8 (1)

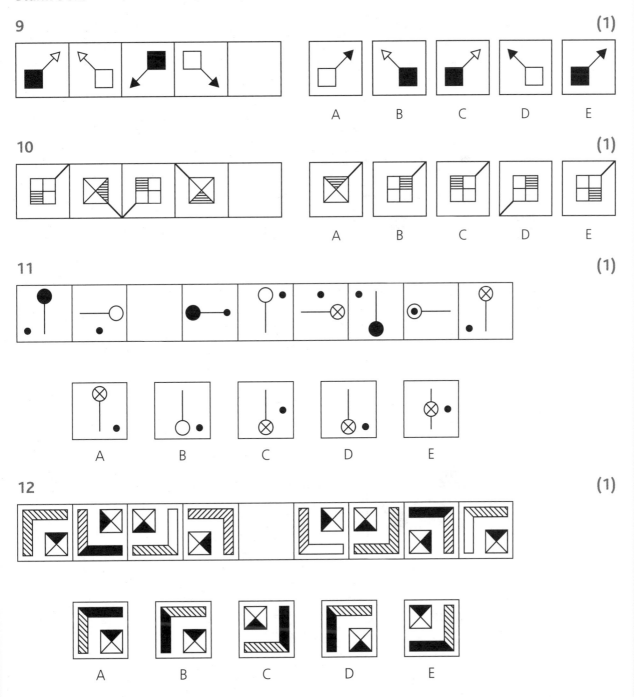

Questions 9–15: The boxes on the left, or above, show a pattern that is arranged in a sequence. Choose the answer option that completes the sequence when inserted in the blank box.

9 (1)

10 (1)

11 (1)

12 (1)

13 (1)

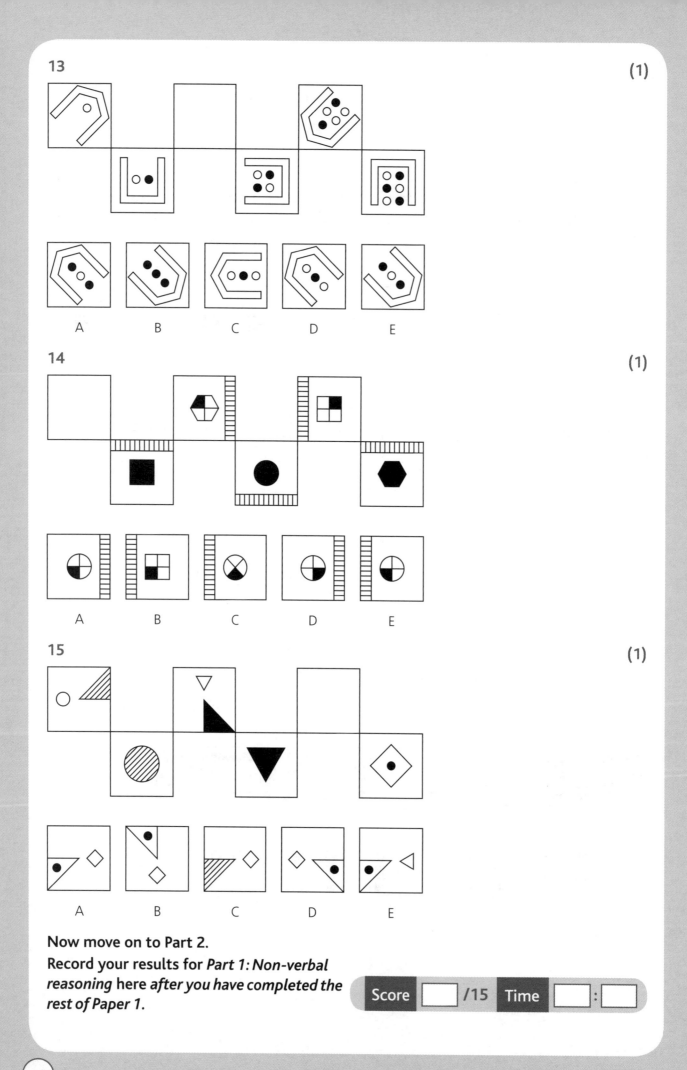

A B C D E

14 (1)

A B C D E

15 (1)

A B C D E

Now move on to Part 2.

Record your results for *Part 1: Non-verbal reasoning* here *after you have completed the rest of Paper 1.*

Score ☐ /15 Time ☐:☐

Part 2: Spatial reasoning

How to answer these questions

All your answers to this part should be recorded on the answer sheet you have downloaded. Look at the examples and then complete the training questions, which have answers at the end of this introduction. **Do not begin timing yourself until you have finished these pages.**

Only check your answers after completing all of Paper 1. The answers are in a cut-out section at the end of the book. Complete the 'results' boxes at the end of this part when you have added up your score. If you run over the time given, complete the questions and note the time you have taken.

Example question 1

Find the cube, or other 3-D shape, that can be made from the net shown on the left.

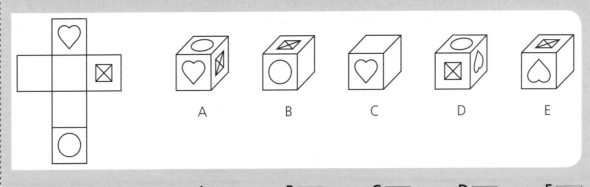

Answer: **A**   When the net is folded to make the cube, the heart will point to a blank face, so options D and E are impossible. When the net is folded, the heart and circle faces will become adjacent faces, with the heart pointing away from the circle. This makes C impossible because the heart is pointing away from a blank face and not the circle. The right-hand face would not be blank if the cube was made up as in B; it would show the heart (pointing away from the circle). A is the only cube that matches the net.

Now answer Training question 1, recording your answer below as shown in the example.

Training question 1

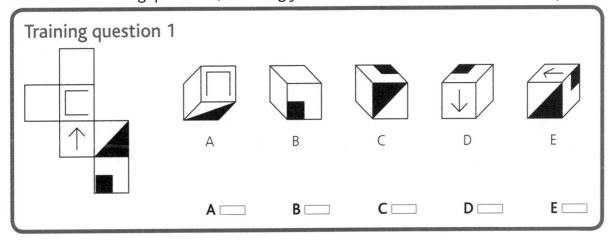

Example question 2

The picture on the left is rotated. Choose the option on the right that represents this picture after it has been rotated.

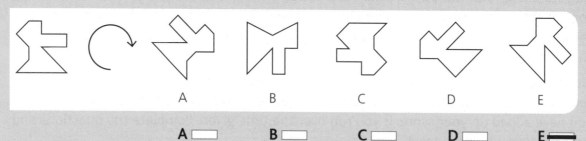

A	B	C	D	E
A ▭	B ▭	C ▭	D ▭	E ▬

Answer: **E**   **rotation** – the picture has been rotated 45° clockwise; **shape** – the shape of the picture does not change.

Now answer Training question 2, recording your answer below as shown in the example.

Training question 2

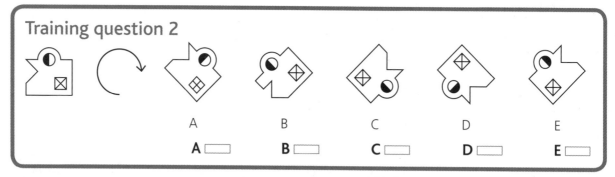

A	B	C	D	E
A ▭	B ▭	C ▭	D ▭	E ▭

Example question 3

One group of separate blocks has been joined together to make the pattern of blocks shown on the far left. Some of the blocks may have been rotated. Select the set of blocks that makes that pattern.

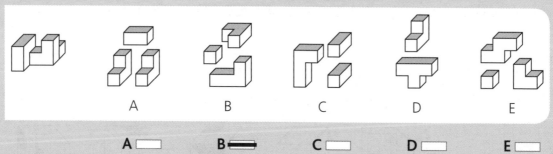

A	B	C	D	E
A ▭	B ▬	C ▭	D ▭	E ▭

Answer: **B**   The short 'L' shape is tipped forward and rotated 90° clockwise. The long 'L' shape slots together with the short 'L' shape. The far left end of the long 'L' shape fits under the short 'L' shape and is hidden from view. The single cube sits behind the long 'L' shape on the right.

Now answer Training question 3, recording your answer below as shown in the example.

Training question 3

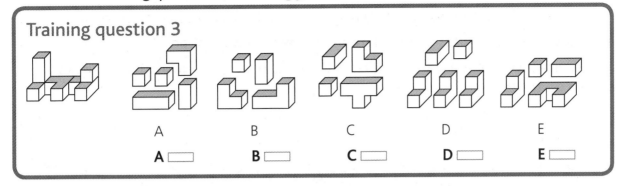

A   B   C   D   E

A ☐   **B** ☐   **C** ☐   **D** ☐   **E** ☐

Training question answers

1C When the net is folded to make a cube, the arrow with the face towards will point towards the three straight lines and will be on the *left* side of these lines. This makes option A impossible as the face on the left would contain the arrow rather than being blank. The arrow will point away from the face with the black square will touch the edge of the face it shares with the face in the top left corner. This makes options D and E impossible. Option B shows the black square in the bottom left corner of the face on the front of the cube; in this configuration, the net clearly shows that the adjacent faces would be the arrow and the face with the black/white triangles. Option C is the only cube that matches the net.

2D rotation – the picture is rotated 225° clockwise; **shading** – the colour/pattern of the small shapes does not change; **shape** – the shape of the picture does not change.

3E The 'L' shape is tipped over to the left and then slots in next to the 'U' shape, on the right side of the 'U'. The single cube sits on top on the 'L'. The cuboid is tipped sideways so that it stands upright; it then sits on top of the 'U' shape on the back row in the far left corner.

Use the downloaded answer sheet to record your answers to the questions that follow.

You will see that the examples and training questions have already been recorded.

You now have 12 minutes to complete the following 22 questions.

Questions 1–6: Find the cube, or other 3-D shape, that can be made from the net shown on the left.

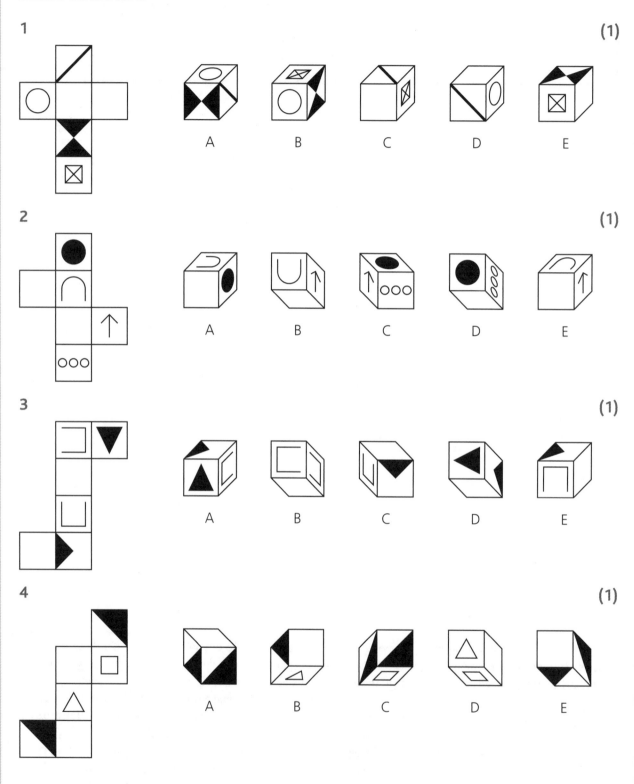

1 (1)

A   B   C   D   E

2 (1)

A   B   C   D   E

3 (1)

A   B   C   D   E

4 (1)

A   B   C   D   E

5 (1)

6 (1)

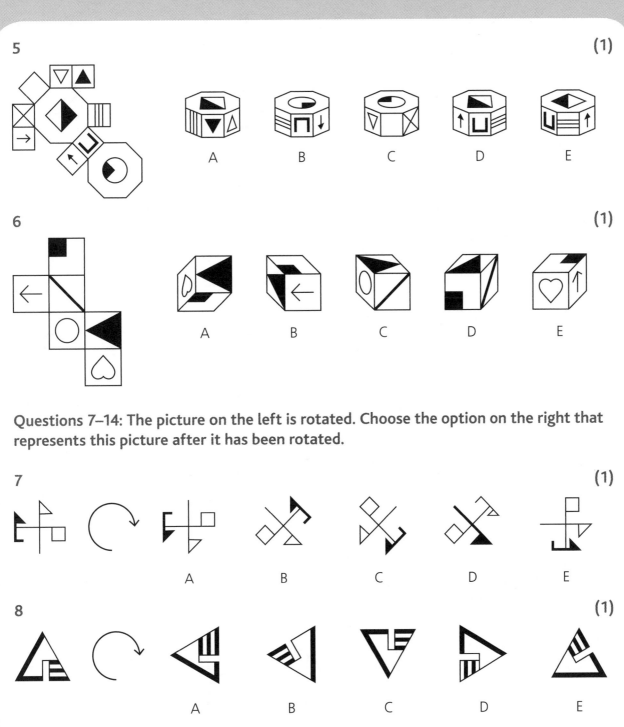

Questions 7–14: The picture on the left is rotated. Choose the option on the right that represents this picture after it has been rotated.

7 (1)

8 (1)

9 (1)

10 (1)

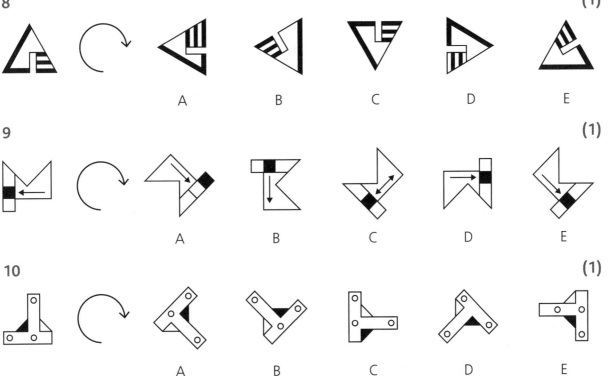

11   (1)

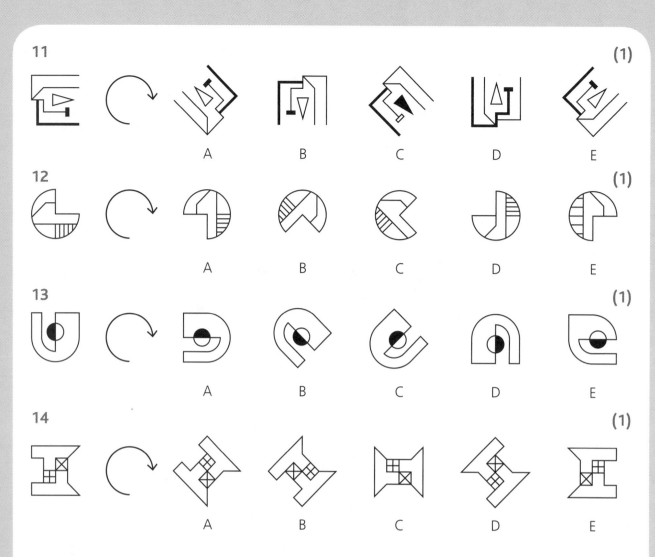

A   B   C   D   E

12   (1)

13   (1)

14   (1)

Questions 15–22: One group of separate blocks has been joined together to make the pattern of blocks shown on the far left. Some of the blocks may have been rotated. Select the set of blocks that makes that pattern.

15   (1)

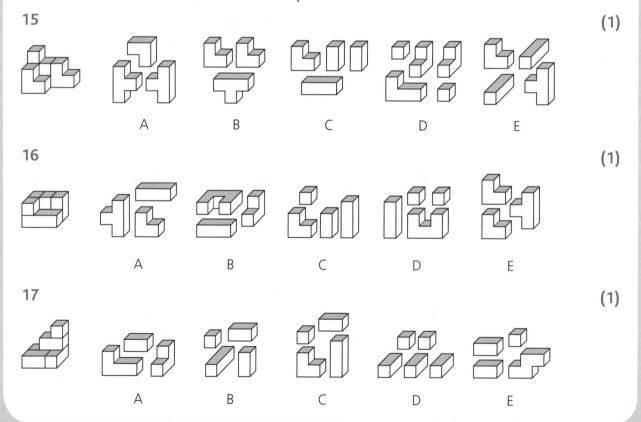

A   B   C   D   E

16   (1)

A   B   C   D   E

17   (1)

A   B   C   D   E

18 (1)

19 (1)

20 (1)

21 (1)

22 (1)

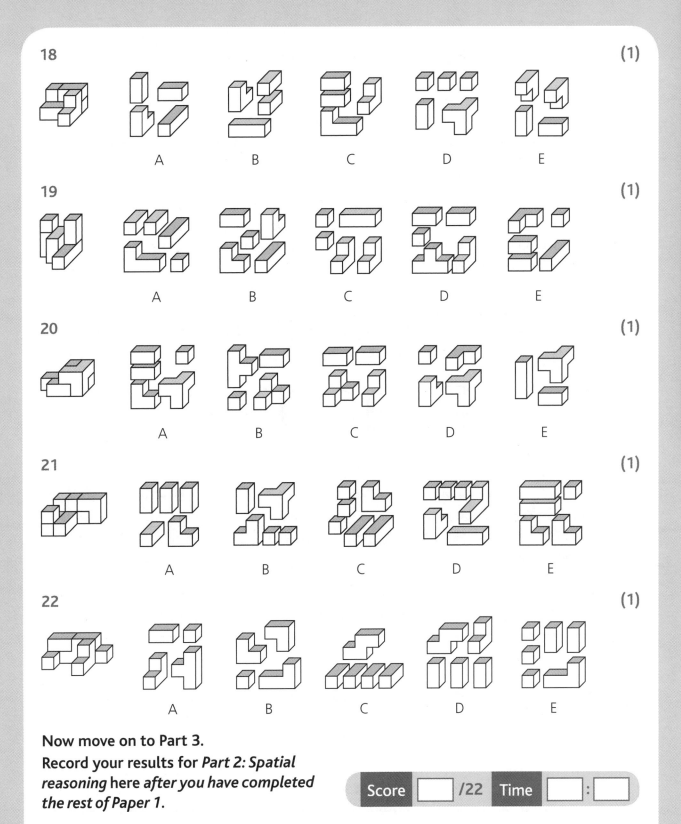

Now move on to Part 3.

Record your results for *Part 2: Spatial reasoning* here *after you have completed the rest of Paper 1*.

Score ☐ /22   Time ☐ : ☐

Part 3: Non-verbal and spatial reasoning

How to answer these questions

All your answers to this part should be recorded on the answer sheet you have downloaded. Look at the examples and then complete the training questions, which have answers at the end of this introduction. **Do not begin timing yourself until you have finished these pages.**

Only check your answers after completing all of Paper 1. The answers are in a cut-out section at the end of the book. Complete the 'results' boxes at the end of this part when you have added up your score. If you run over the time given, complete the questions and note the time you have taken.

Example question 1

The picture on the left is reflected in a vertical mirror line. Select the option on the right that represents the picture after it has been reflected.

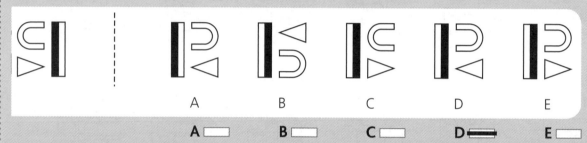

Answer: **D** The complete picture is reflected in a vertical mirror line so the whole picture flips to the right. Note that the point of the triangle and the open end of the arch shape are next to the black half of the rectangle in the original picture. These features must still be the same in the reflected picture.

Now answer Training question 1, recording your answer below as shown in the example.

Training question 1

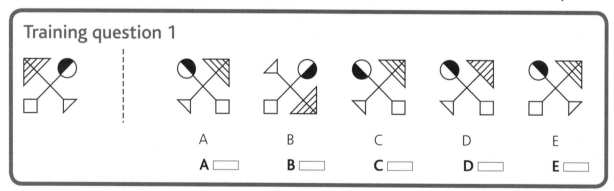

Example question 2

One of the boxes on the right completes the pattern in the grid on the left. Select the option that completes the pattern.

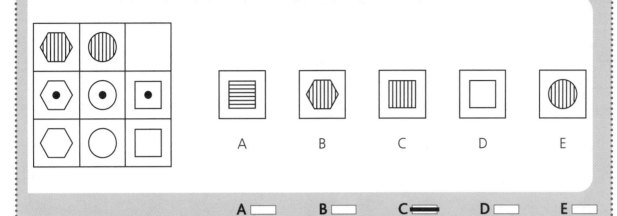

Answer: **C**   **shape** – shapes match vertically; **shading** – colours/patterns match horizontally.

Now answer Training question 2, recording your answer below as shown in the example.

Training question 2

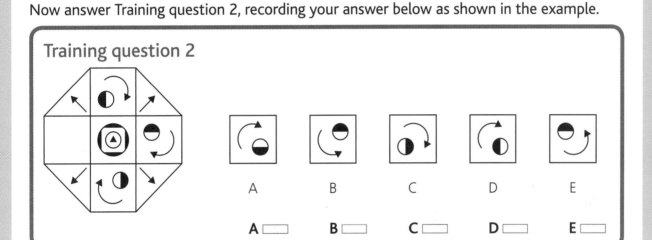

Training question answers

1A The complete picture is reflected in a vertical mirror line so the whole picture flips to the right. Note that option B has been reflected in a diagonal mirror line; the shading of the circle in C is incorrect; the centre line visible in the original triangle is no longer evident in the reflected picture in D; the bottom shapes have not been reflected in option E.

2A symmetry – the four triangular areas of the grid are symmetrical both horizontally and vertically; **rotation** – the image in each square rotates 90° clockwise each time.

Use the downloaded answer sheet to record your answers to the questions that follow.

You will see that the examples and training questions have already been recorded.

You now have 6 minutes to complete the following 12 questions.

Questions 1–7: The picture on the left is reflected in a vertical mirror line. Select the option on the right that represents the picture after it has been reflected.

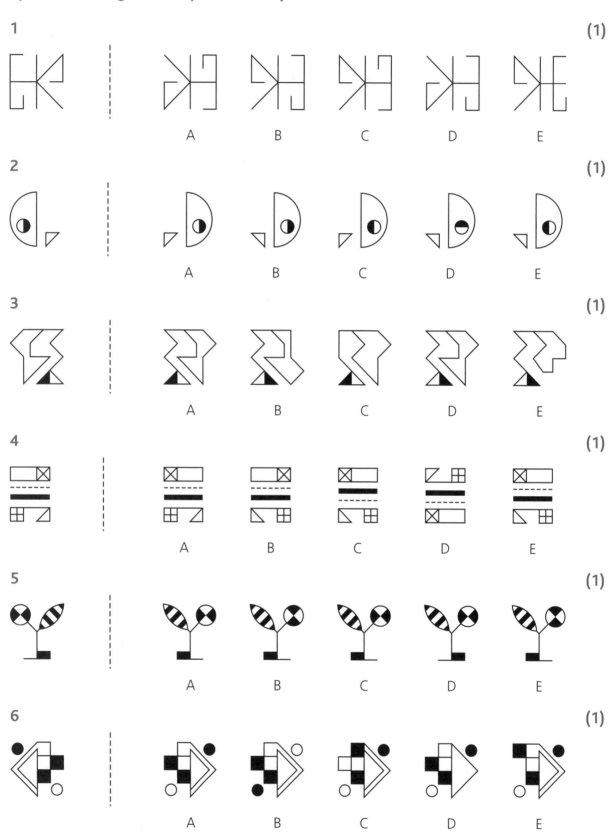

1 (1)

A   B   C   D   E

2 (1)

A   B   C   D   E

3 (1)

A   B   C   D   E

4 (1)

A   B   C   D   E

5 (1)

A   B   C   D   E

6 (1)

A   B   C   D   E

7 (1)

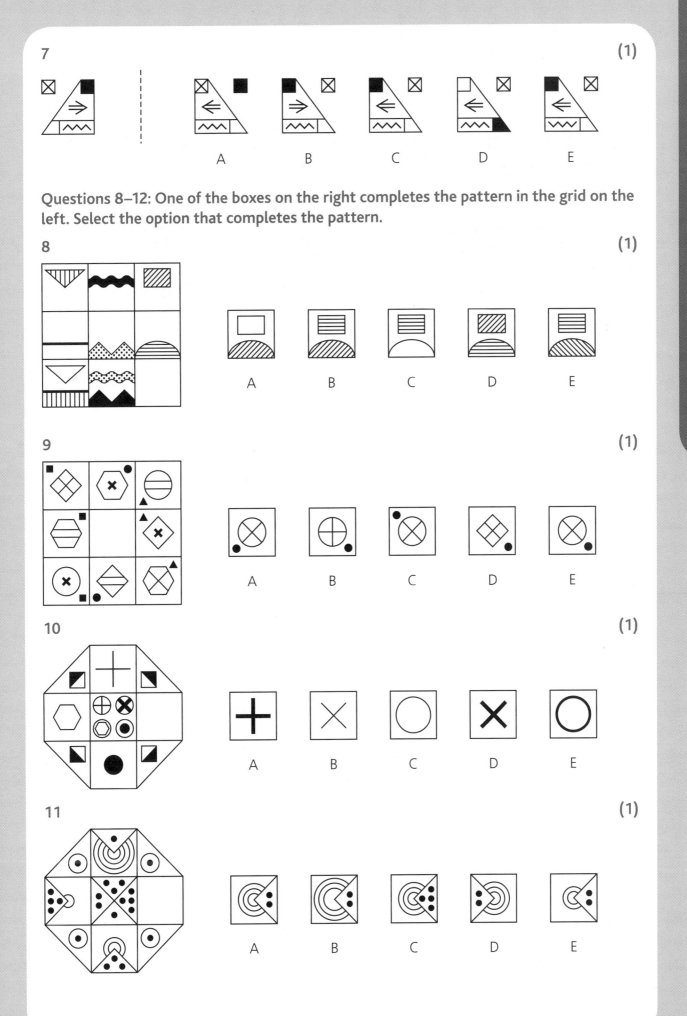

Questions 8–12: One of the boxes on the right completes the pattern in the grid on the left. Select the option that completes the pattern.

8 (1)

9 (1)

10 (1)

11 (1)

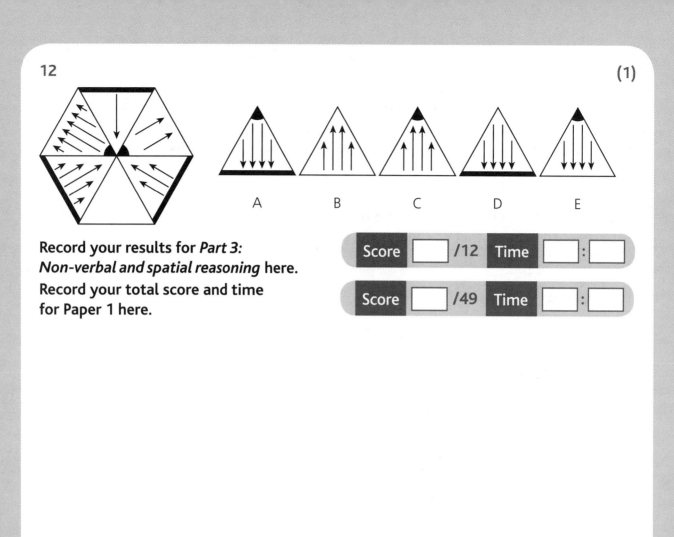

A   B   C   D   E

Record your results for *Part 3:*
Non-verbal and spatial reasoning here.
Record your total score and time
for Paper 1 here.

Score	/12	Time	:

Score	/49	Time	:

● Paper 2

Download and print the answer sheet from galorepark.co.uk/answersheets before you start this paper.

Complete all three parts of this paper according to the timings given at the start of each set of questions. Stop the timer after completing each part and start it again after answering the training questions.

Part 1: Spatial reasoning

How to answer these questions

All your answers to this part should be recorded on the answer sheet you have downloaded. Look at the example. **Do not begin timing yourself until you have finished this page.**

Only check your answers after completing all of Paper 2. The answers are in a cut-out section at the end of the book. Complete the 'results' boxes at the end of this part when you have added up your score. If you run over the time given, complete the questions and note the time you have taken.

Example question

In Questions 1–6 you will see a rotated version of each of the 3-D diagrams shown (A–F) and you will need to match each diagram to its rotated image. In this example question, diagram 1 is a rotated version of one of the 3-D diagrams shown (A or B). Choose the diagram that matches.

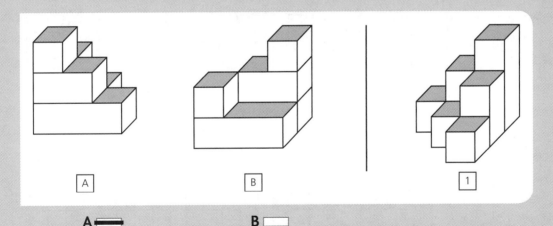

Answer: **A**   Diagram A is tipped to its left and then rotated 90º anticlockwise around a vertical line.

Use the downloaded answer sheet to record your answers to the questions that follow.

You will see that the example has already been recorded.

You now have 7 minutes to complete the following 12 questions.

Questions 1–6: Diagrams 1–6 are rotated versions of the 3-D diagrams shown in A–F. Match each 3-D diagram to its rotated image.

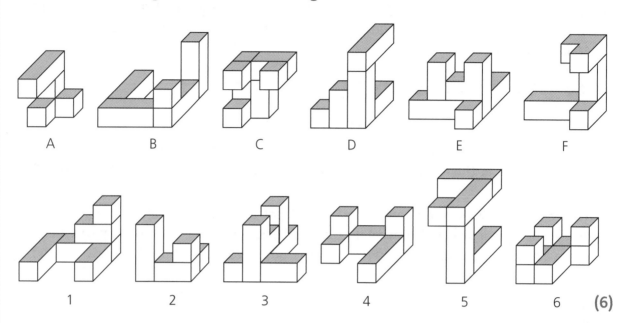

(6)

Questions 7–12: Which of the answer options is a 2-D plan of the 3-D diagram on the left, when viewed from above?

7 (1)

8 (1)

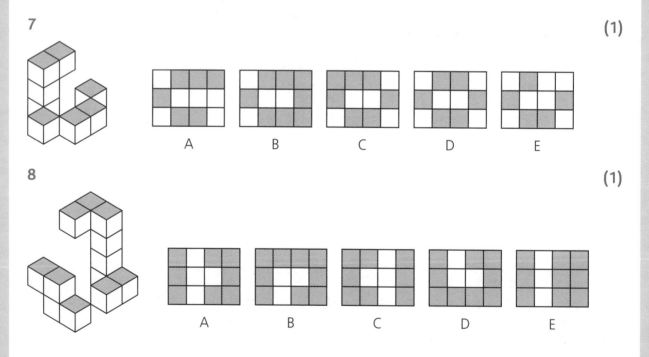

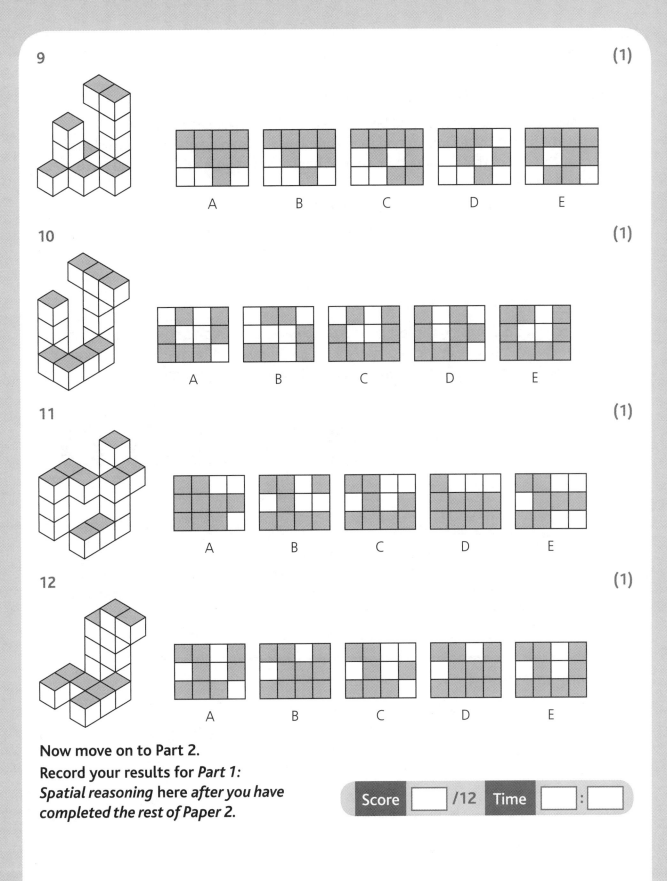

9 (1)

10 (1)

11 (1)

12 (1)

Now move on to Part 2.

Record your results for *Part 1: Spatial reasoning* here *after you have completed the rest of Paper 2.*

Score [ ] /12 Time [ ] : [ ]

Part 2: Non-verbal and spatial reasoning

How to answer these questions

All your answers to this paper should be recorded on the answer sheet you have downloaded. Look at the examples and then complete the training questions, which have answers at the end of this introduction. **Do not begin timing yourself until you have finished these pages.**

Only check your answers after completing all of Paper 2. The answers are in a cut-out section at the end of the book. Complete the 'results' boxes at the end of this part when you have added up your score. If you run over the time given, complete the questions and note the time you have taken.

Example question 1

Look at the two pictures on the left connected by an arrow. Decide how the first picture has been changed to create the second. Now apply the same rules to the third picture and choose how it has been changed from the five options on the right.

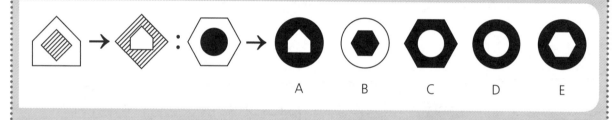

Answer: **E**   **size/position** – the large shape and the small shape swap places: the large shape decreases in size to become the inner shape and the small shape increases in size to become the outer shape; **shading/pattern** – both shapes retain their shading.

Now answer Training question 1, recording your answer below as shown in the example.

Training question 1

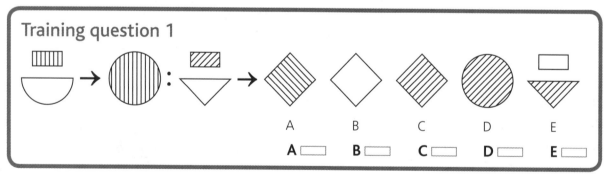

Example question 2

Look at the face given on the left. When the net is folded into a cube, which face will appear opposite it?

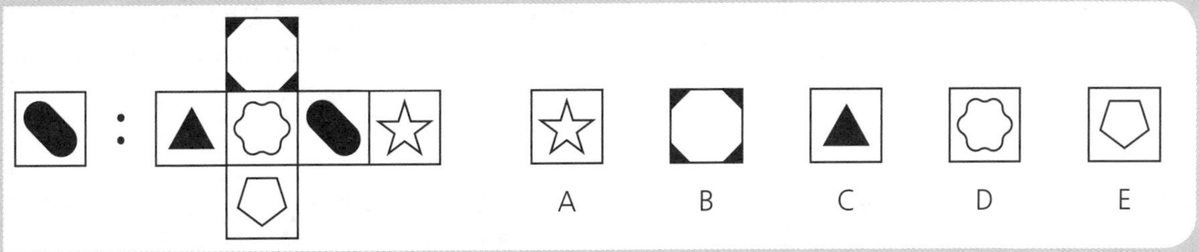

A ☐ B ☐ C ☒ D ☐ E ☐

Answer: **C** When the net is folded to make the cube, the face with the black triangle will be opposite the target face (black lozenge). Imagine folding together the face with the black lozenge and the face with the white pentagon, so that the face with the white wavy shape becomes the top face. Then imagine folding down the face with the four black corner triangles. When you fold down the face with the black triangle, it will be the opposite face to the target face.

Now answer Training question 2, recording your answer below as shown in the example.

Training question 2

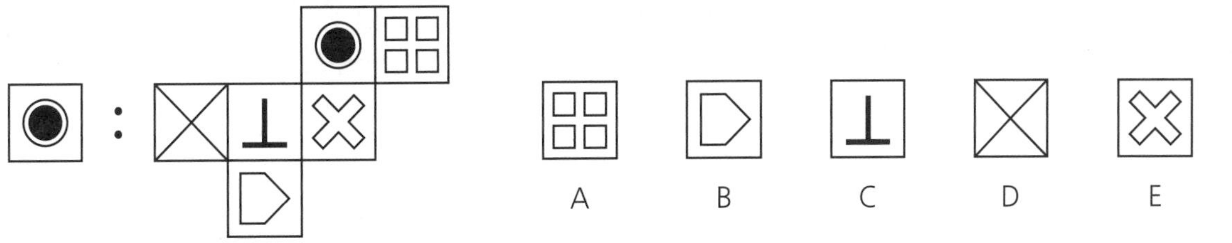

A ☐ B ☐ C ☐ D ☐ E ☐

Training question answers

1C **shape/reflection** – the shape at the base of the picture is duplicated and reflects horizontally to form a new whole shape, while the rectangle is removed; **shading/pattern** – the new shape is shaded in the same way as the original rectangle.

2B When the net is folded to make the cube, the face with the irregular pentagon will be opposite the target face (circles). Imagine folding together the face with the circle and the face with the 'T' shape (so that the face with the white solid cross becomes the top face). Then imagine folding round the face with the irregular pentagon; it will sit opposite the target face.

Use the downloaded answer sheet to record your answers to the questions that follow.

You will see that the examples and training questions have already been recorded.

You now have 7 minutes to complete the following 13 questions.

Questions 1–7: Look at the two pictures on the left connected by an arrow. Decide how the first picture has been changed to create the second. Now apply the same rules to the third picture and choose how it has been changed from the five options on the right.

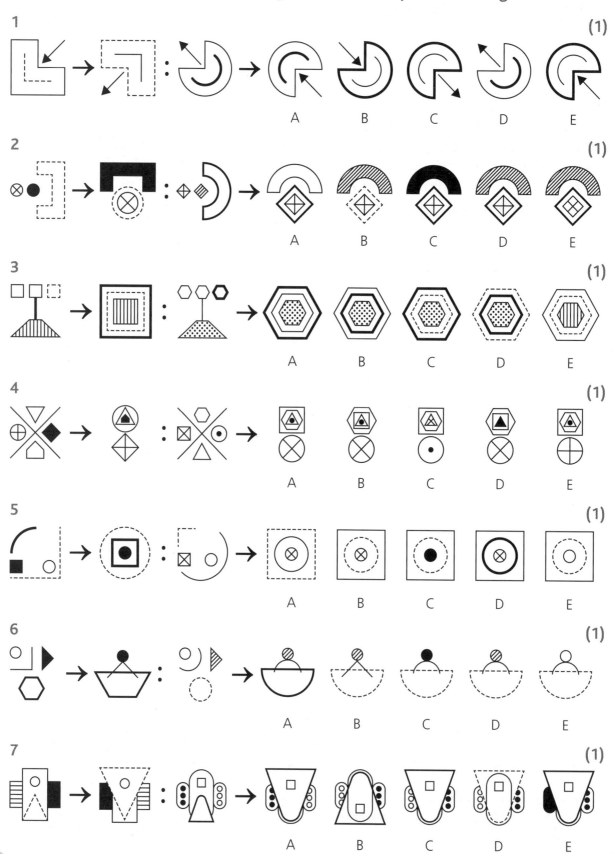

1 (1)

A B C D E

2 (1)

A B C D E

3 (1)

A B C D E

4 (1)

A B C D E

5 (1)

A B C D E

6 (1)

A B C D E

7 (1)

A B C D E

Questions 8–13: Look at the face given on the left. When the net is folded into a cube, which face will appear opposite it?

8 (1)

A B C D E

9 (1)

A B C D E

10 (1)

A B C D E

11 (1)

A B C D E

12 (1)

A B C D E

13 (1)

A B C D E

Now move on to Part 3.
Record your results for *Part 2: Non-verbal and spatial reasoning* here *after you have completed the rest of Paper 2.*

Score ☐ /13   Time ☐ : ☐

Part 3: Non-verbal reasoning

How to answer these questions

All your answers to this paper should be recorded on the answer sheet you have downloaded. Where there are examples, look at these and then complete the training questions, which have answers at the end of this introduction. **Do not begin timing yourself until you have finished these pages.**

Only check your answers after completing all of Paper 2. The answers are in a cut-out section at the end of the book. Complete the 'results' boxes at the end of this part when you have added up your score. If you run over the time given, complete the questions and note the time you have taken.

Example question 1

Look at this set of pictures. Identify the one that is *most unlike* the others.

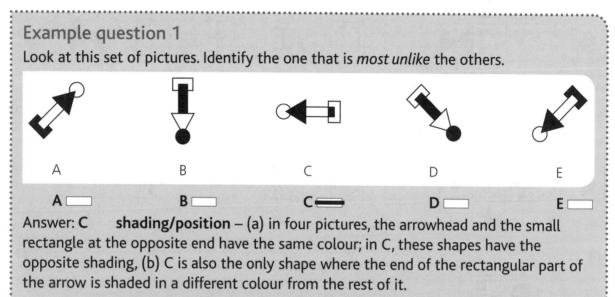

Answer: **C**   shading/position – (a) in four pictures, the arrowhead and the small rectangle at the opposite end have the same colour; in C, these shapes have the opposite shading, (b) C is also the only shape where the end of the rectangular part of the arrow is shaded in a different colour from the rest of it.

Distractor: **direction** – the direction in which the picture is pointing is unimportant.

Now answer Training question 1, recording your answer below as shown in the example.

Training question 1

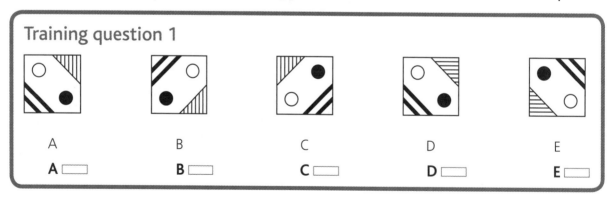

Example question 2

Look at the first two pictures and decide what they have in common. Then select the option from the five on the right that belongs to the same set.

A   B   C   D   E

A ▭   B ▭   C ▭   D ▭   E ▬

Answer: **E**   **shading** – the small inner shape is shaded with a diagonal striped pattern going from top right to bottom left.

Now answer Training question 2, recording your answer below as shown in the example.

Training question 2

A   B   C   D   E

A ▭   B ▭   C ▭   D ▭   E ▭

Example question 3

Each letter represents an individual feature in the picture to its left. Work out which feature is represented by each letter. Then apply the rules to the picture in the box and select the code that fits it.

  RF

  SG   RG   SF   TF   TG   SG

  TF   A   B   C   D   E

A ▭   B ▬   C ▭   D ▭   E ▭

Answer: **B**   **shape** – the first letter represents shape: **R** is a circle, **S** is a square and **T** is a triangle; **shading** – the second letter represents shading: **F** is diagonal stripes and **G** is plain.

Now answer Training question 3, recording your answer below as shown in the example.

Training question 3

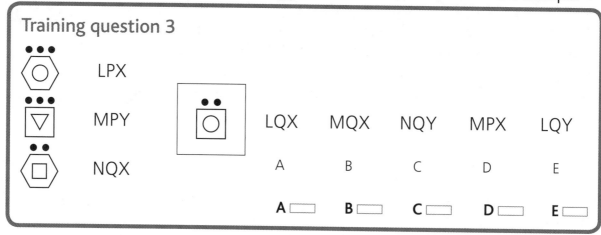

LQX	MQX	NQY	MPX	LQY
A	B	C	D	E

A ☐ B ☐ C ☐ D ☐ E ☐

Training question answers

1A **shading** – if you were to extend the striped shading in the corner of the square it would cover the white circle in all options except A because A has been flipped instead of rotated.

2B **number/size** – each picture contains three shapes: a large shape, a medium-sized shape and a small shape;
shape/proportion – each picture contains two identical shapes, one of which is half the size of the other.
Distractors: **position** – the position of the smaller shapes (inside/outside the large shape) is unimportant;
shading – the small shape can be shaded or unshaded.

3E **shape** – (a) the first letter represents the small inner shape: L is a circle, M is a triangle and N is a square,
(b) the third letter represents the large shape: X is a hexagon and Y is a square; **number** – the second letter represents the number of circles above the large shape: **P** is three, while **Q** is two.

34

Use the downloaded answer sheet to record your answers to the questions that follow.

You will see that the examples and training questions have already been recorded.

You now have 11 minutes to complete the following 24 questions.

Questions 1–6: Look at this set of pictures. Identify the one that is *most unlike* the others.

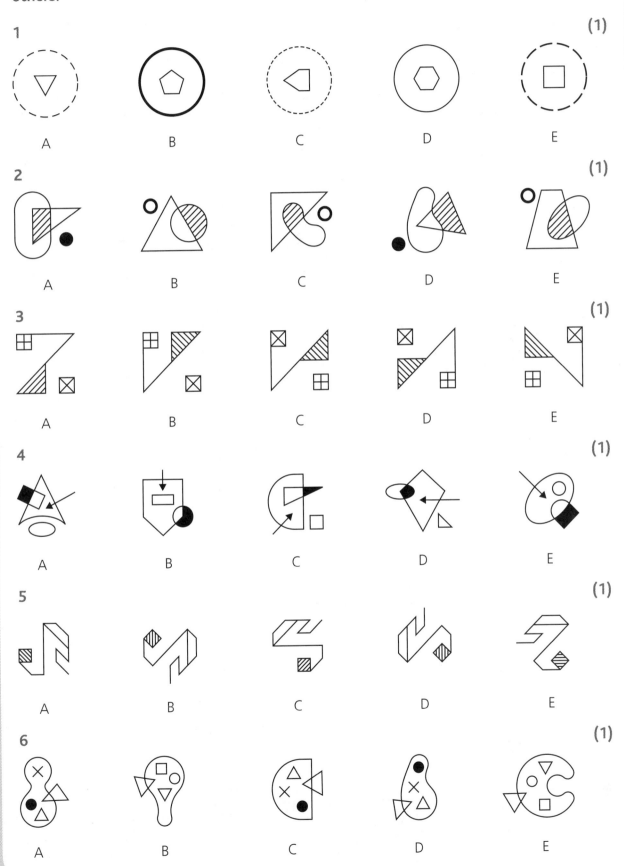

1   (1)

A   B   C   D   E

2   (1)

A   B   C   D   E

3   (1)

A   B   C   D   E

4   (1)

A   B   C   D   E

5   (1)

A   B   C   D   E

6   (1)

A   B   C   D   E

Questions 7–16: Look at the first two pictures and decide what they have in common. Then select the option from the five on the right that belongs to the same set.

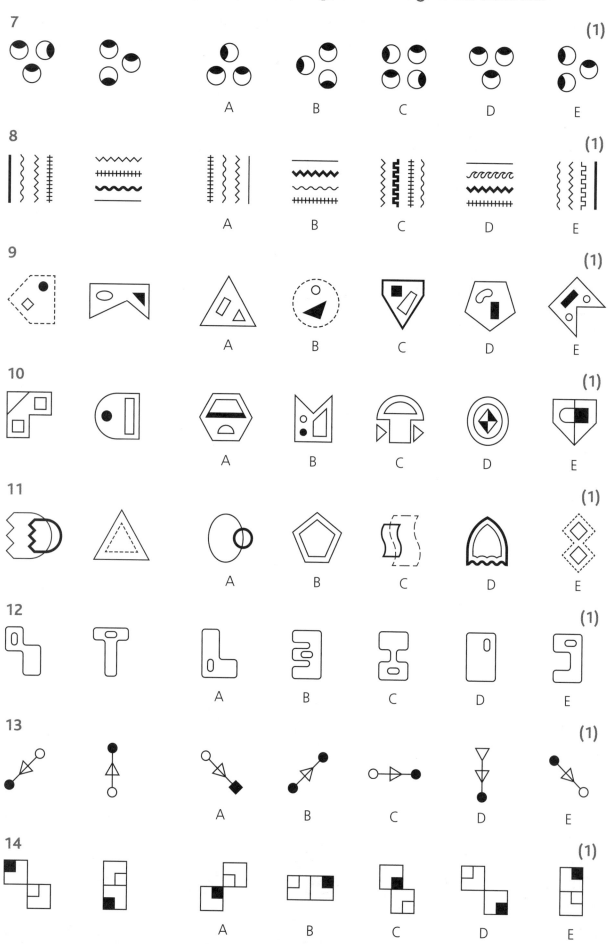

7 (1)

A B C D E

8 (1)

A B C D E

9 (1)

A B C D E

10 (1)

A B C D E

11 (1)

A B C D E

12 (1)

A B C D E

13 (1)

A B C D E

14 (1)

A B C D E

15 (1)

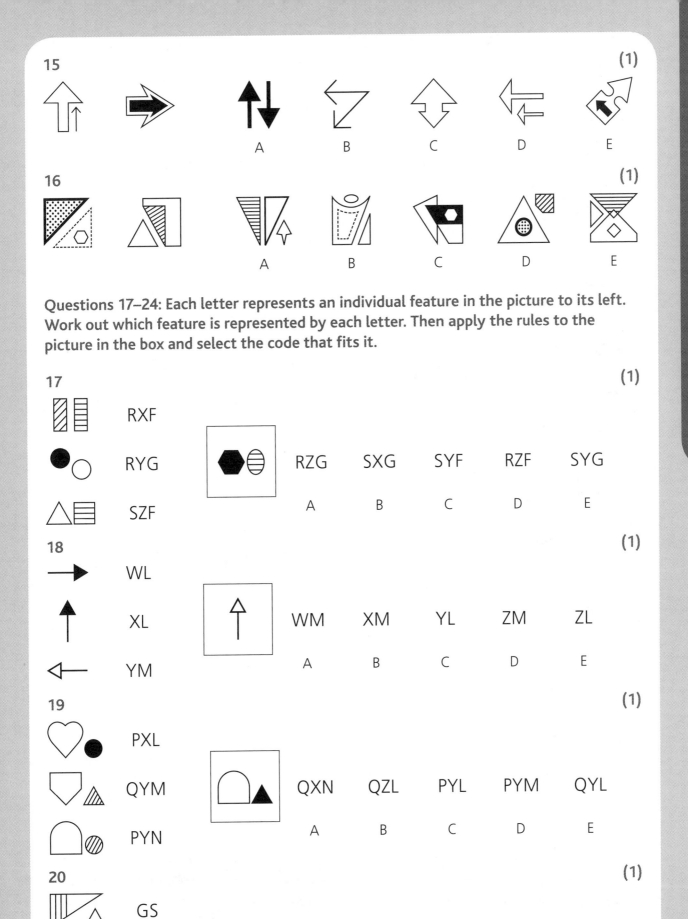

A   B   C   D   E

16 (1)

A   B   C   D   E

Questions 17–24: Each letter represents an individual feature in the picture to its left. Work out which feature is represented by each letter. Then apply the rules to the picture in the box and select the code that fits it.

17 (1)

RXF

RYG

SZF

	RZG	SXG	SYF	RZF	SYG
	A	B	C	D	E

18 (1)

WL

XL

YM

	WM	XM	YL	ZM	ZL
	A	B	C	D	E

19 (1)

PXL

QYM

PYN

	QXN	QZL	PYL	PYM	QYL
	A	B	C	D	E

20 (1)

GS

HS

GT

	GS	HS	GT	HT	FT
	A	B	C	D	E

21 (1)

	PX
	QY
	RZ
	PY

QX   RY   RX   QZ   PZ

A   B   C   D   E

22 (1)

	LR
	MS
	NR
	MT

LS   MR   LT   NS   NT

A   B   C   D   E

23 (1)

	RFL
	SGM
	RHN
	TGL

SFN   THM   SHL   RGL   TFN

A   B   C   D   E

24 (1)

	XPF
	YQG
	ZRF
	XSH

YRH   XQF   ZSG   YPH   ZQF

A   B   C   D   E

Record your results for *Part 3: Non-verbal reasoning* here.

Record your total score and time for Paper 2 here.

Score ☐ /24   Time ☐ : ☐

Score ☐ /49   Time ☐ : ☐

● Paper 3

Complete all three parts of this paper according to the timings given at the start of each set of questions. Stop the timer after completing each part and start it again after answering the training questions for the next part.

Part 1: Non-verbal and spatial reasoning

How to answer these questions

All your answers should be recorded on this paper. Record the letter for the answer you have chosen on the line provided. Look at the examples and then complete the training questions, which have answers at the end of this introduction. **Do not begin timing yourself until you have finished these pages.**

Only check your answers after completing all of Paper 3. The answers are in a cut-out section at the end of the book. Complete the 'results' boxes at the end of this part when you have added up your score. If you run over the time given, complete the questions and note the time you have taken.

Example question 1

One group of separate blocks has been joined together to make the pattern of blocks shown on the far left. Some of the blocks may have been rotated. Find the set of blocks that makes that pattern.

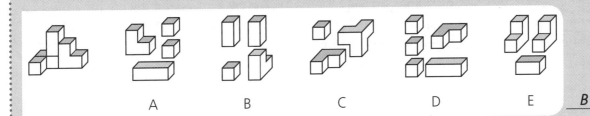

A   B   C   D   E   _B_

Answer: **B**   The 'L' shape is rotated 90° clockwise. One of the cuboids is tipped sideways so that it sits on its long edge. This cuboid joins together with the 'L' shape, to the left of the 'L' on the bottom layer of the model. The remaining cuboid sits on top of the first cuboid, immediately adjacent to the 'L' shape. The single cube sits in front of the cuboid on the bottom layer of the model, on the far left.

Now answer Training question 1, recording your answer choice as shown in the example.

Training question 1

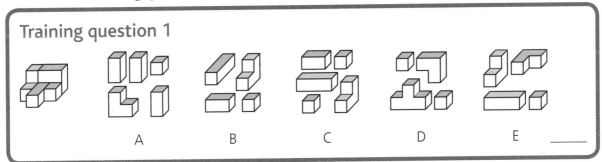

A   B   C   D   E   ____

Example question 2

Look at the first three pictures and decide what they have in common. Then find the option from the five on the right that belongs to the same set.

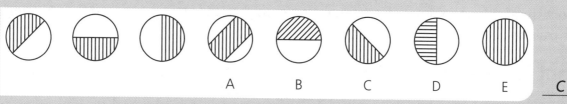

Answer: **C**   **shading/proportion** – each picture has exactly one-half shaded; **direction** – the striped shading is vertical.

Now answer Training question 2, recording your answer choice as shown in the example.

Training question 2

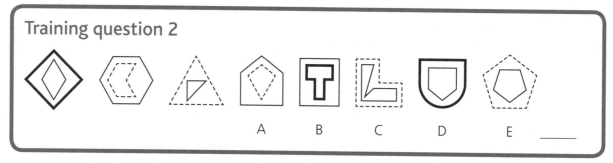

Example question 3

Look at the two sets of shapes on the left connected by arrows and decide how the first shapes have been changed to create the second ones. Now apply the same rules to the third picture and choose the solution from the five options on the right.

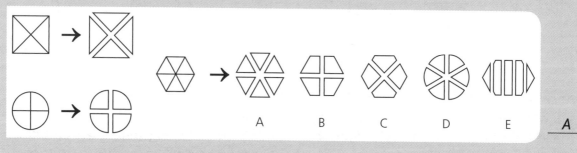

Answer: **A**   **shape/proportion** – the shape in the first picture is divided up into equal segments; the segments in the first shape become separated from the 'whole' in the second picture; **position** – the segments move slightly outwards from the centre of the original shape in the second picture but they do not change size or shape.

Now answer Training question 3, recording your answer choice as shown in the example.

Training question 3

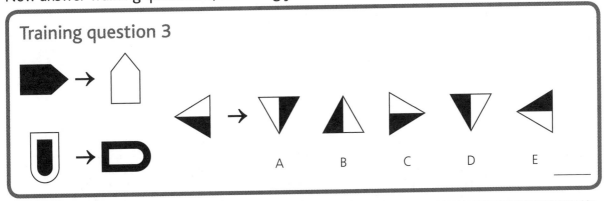

A   B   C   D   E   ____

Example question 4

One of the boxes on the right completes the pattern in the grid on the left. Find the option that completes the pattern.

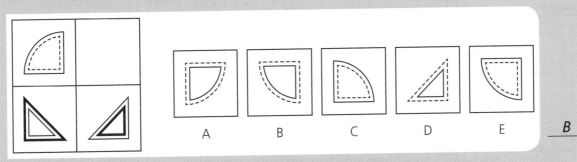

A   B   C   D   E   _B_

Answer: **B**   **shape** – shapes match in a horizontal pattern; **rotation** – shapes rotate 90° anticlockwise in a horizontal pattern, working from left to right; **line style** – line style swaps between inside and outside shapes in a horizontal pattern.

Now answer Training question 4, recording your answer choice as shown in the example.

Training question 4

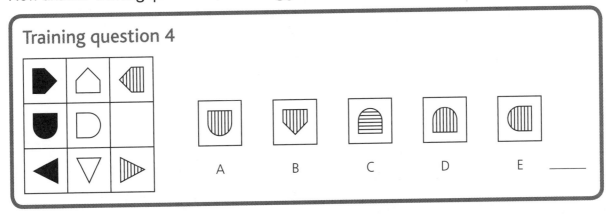

A   B   C   D   E   ____

Training question answers

1E The cuboid is rotated 90° and fits alongside the upright 'L' shape, on the right side of the 'L'. The remaining 'L' shape is tipped forward then rotated 180°. This 'L' shape then slots together with the cuboid and the other 'L' shape on the right side of these shapes. The single cube sits in front of the 'L' shape on the right side of the model.

2E number – the inner shape has the same number of sides as the outer shape (sides must be straight) *Distractor*: **line style** – the style of line is irrelevant.

3D rotation – the shape rotates 90° anticlockwise; **shading** – shaded areas switch to the opposite colour in the second picture.

4D shading – colours/patterns match in a vertical pattern; **shape** – shapes match in a horizontal pattern; **rotation** – working in a horizontal pattern from left to right, shapes rotate 90° anticlockwise.

Record the letter for the answer you have chosen on the line provided.

You now have 10 minutes to complete the following 24 questions.

Questions 1–6: One group of separate blocks has been joined together to make the pattern of blocks shown on the far left. Some of the blocks may have been rotated. Find the set of blocks that makes that pattern.

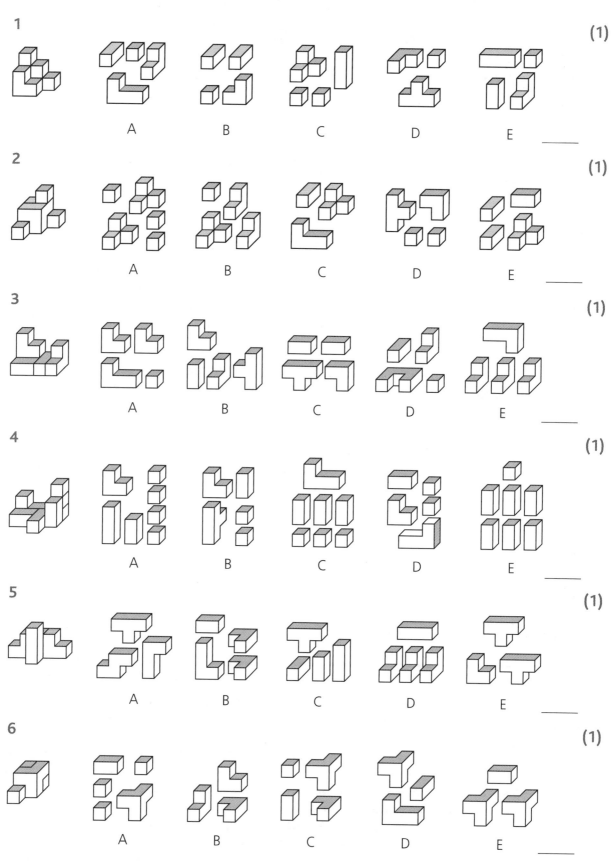

1 (1)

A B C D E ____

2 (1)

A B C D E ____

3 (1)

A B C D E ____

4 (1)

A B C D E ____

5 (1)

A B C D E ____

6 (1)

A B C D E ____

Questions 7–14: Look at the first three pictures and decide what they have in common. Then find the option from the five on the right that belongs to the same set.

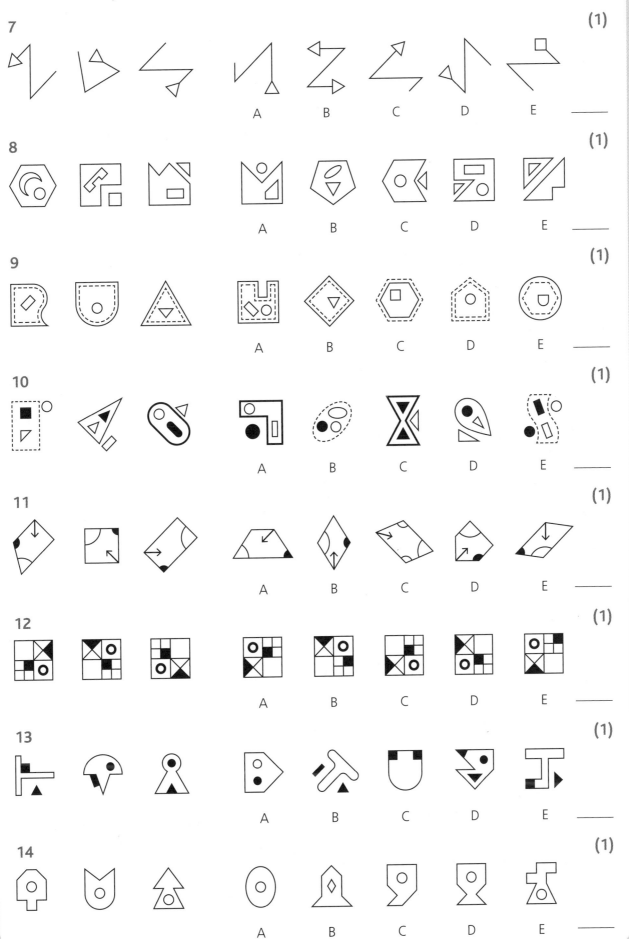

7   (1)

A   B   C   D   E   ——

8   (1)

A   B   C   D   E   ——

9   (1)

A   B   C   D   E   ——

10   (1)

A   B   C   D   E   ——

11   (1)

A   B   C   D   E   ——

12   (1)

A   B   C   D   E   ——

13   (1)

A   B   C   D   E   ——

14   (1)

A   B   C   D   E   ——

Questions 15–19: Look at the two sets of shapes on the left connected by arrows and decide how the first shapes have been changed to create the second. Now apply the same rules to the third picture and choose the solution from the five options on the right.

15　(1)

16　(1)

17　(1)

18　(1)

19　(1)

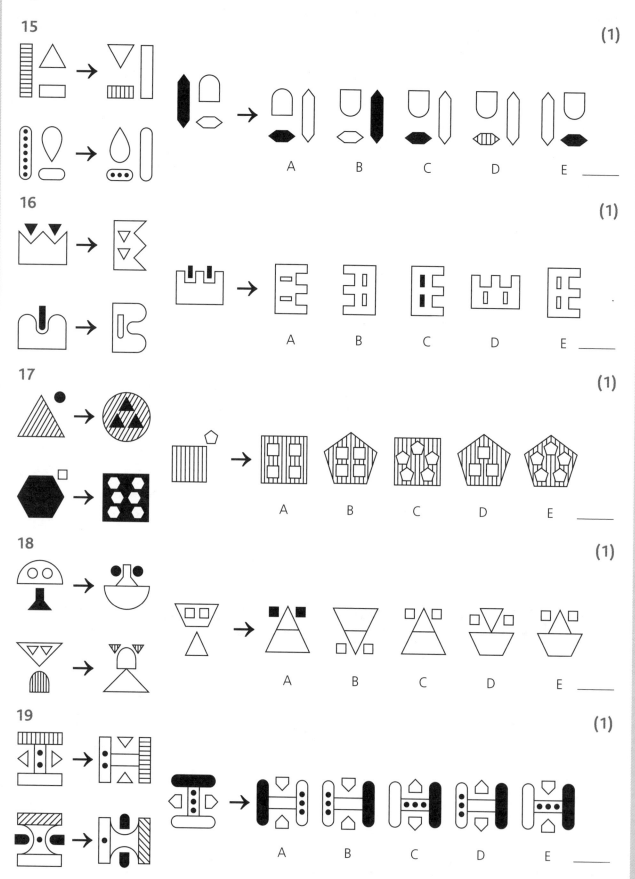

Questions 20–24: One of the boxes on the right completes the pattern in the grid on the left. Find the option that completes the pattern.

20 (1)

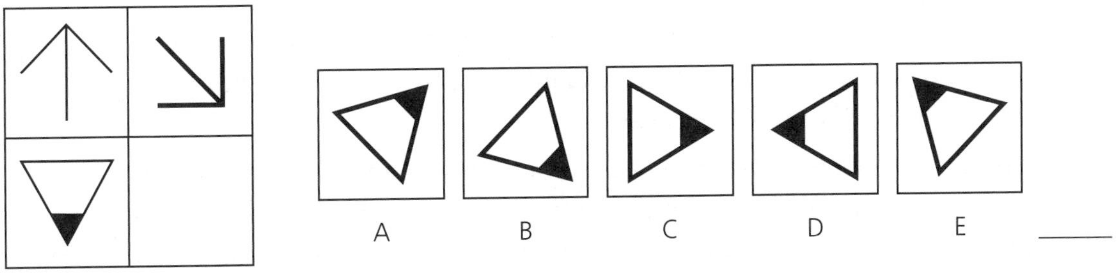

A   B   C   D   E   _____

21 (1)

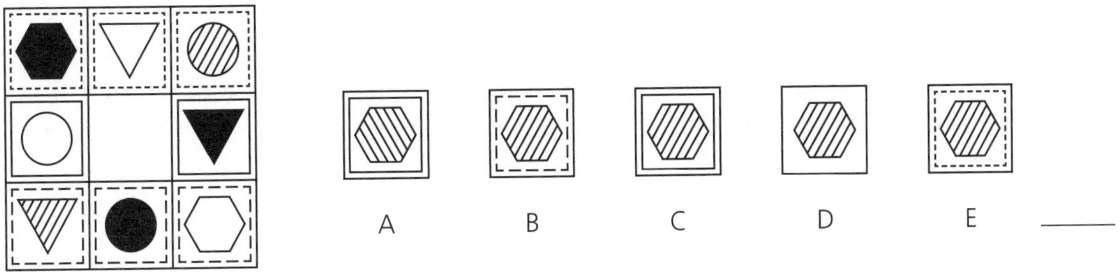

A   B   C   D   E   _____

22 (1)

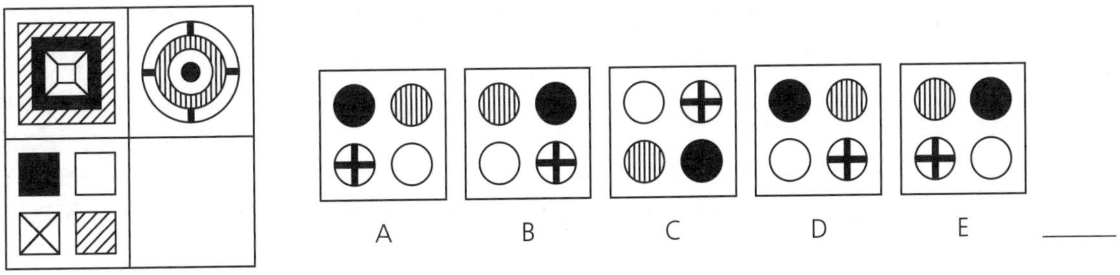

A   B   C   D   E   _____

23 (1)

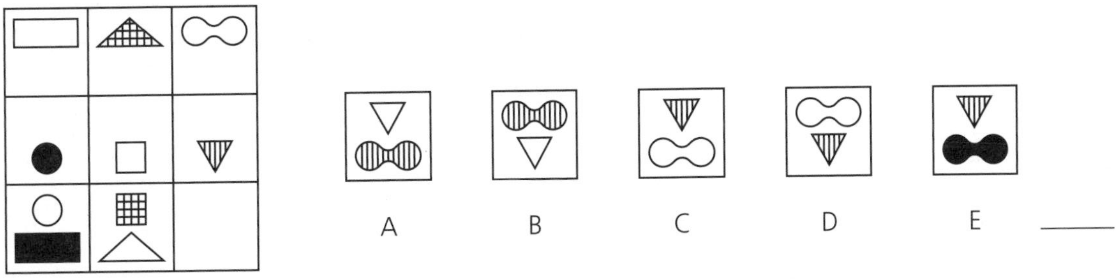

A   B   C   D   E   _____

24 (1)

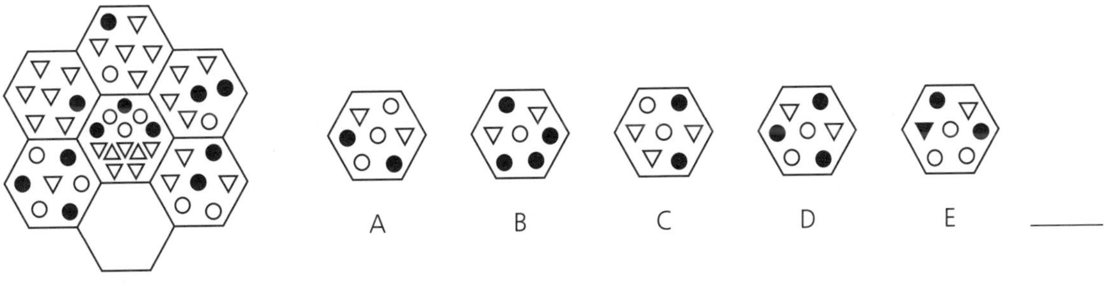

A   B   C   D   E   _____

Now move on to Part 2.

Record your results for *Part 1: Non-verbal and spatial reasoning* here *after you have completed the rest of Paper 3.*

Score [ ] /24   Time [ ] : [ ]

45

Part 2: Non-verbal reasoning

How to answer these questions

All your answers should be recorded on this paper. Record the letter for the answer you have chosen on the line provided. Look at the examples and then complete the training questions, which have answers at the end of this introduction. **Do not begin timing yourself until you have finished these pages.**

Only check your answers after completing all of Paper 3. The answers are in a cut-out section at the end of the book. Complete the 'results' boxes at the end of this part when you have added up your score. If you run over the time given, complete the questions and note the time you have taken.

Example question 1

Each letter represents an individual feature in the picture to its left. Work out which feature is represented by each letter. Then apply the rules to the picture in the box and find the code that fits it.

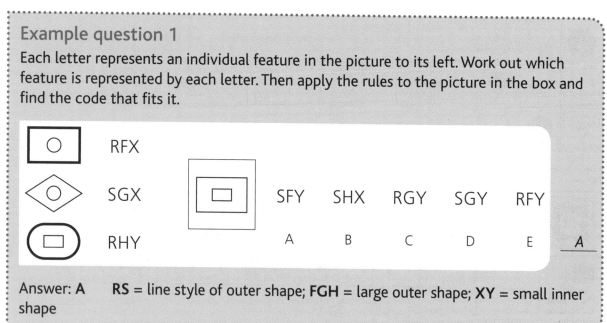

Answer: **A**   **RS** = line style of outer shape; **FGH** = large outer shape; **XY** = small inner shape

Now answer Training question 1, recording your answer choice as shown in the example.

Training question 1

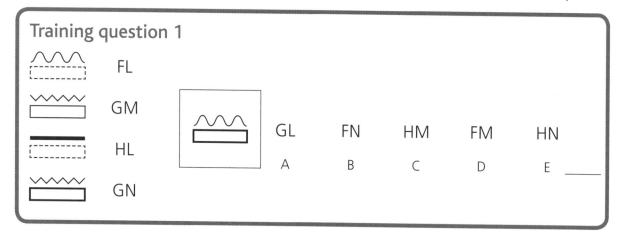

Example question 2

The boxes on the left show a pattern that is arranged in a sequence. Find the answer option that completes the sequence when inserted in the blank box.

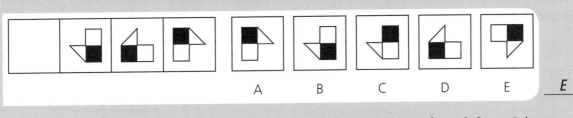

Answer: **E**   **rotation** – the entire picture rotates 90° clockwise from left to right.

Now answer Training question 2, recording your answer choice as shown in the example.

Training question 2

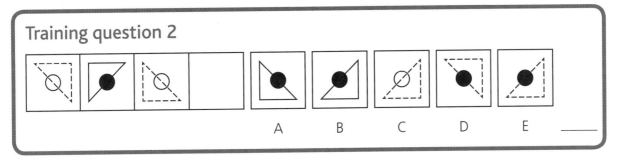

Training question answers

1B FCH = style of upper line; **LMN =** line style of rectangle

2B direction – the triangle rotates 90° anticlockwise; **line style** – the line style of the triangle alternates between solid and dashed; **shading** – the circle alternates between black and white.

Record the letter for the answer you have chosen on the line provided.

You now have 5 minutes to complete the following 12 questions.

Questions 1–5: Each letter represents an individual feature in the picture next to it. Work out which feature is represented by each letter. Apply the rules to the picture in the box and then find the code that fits it.

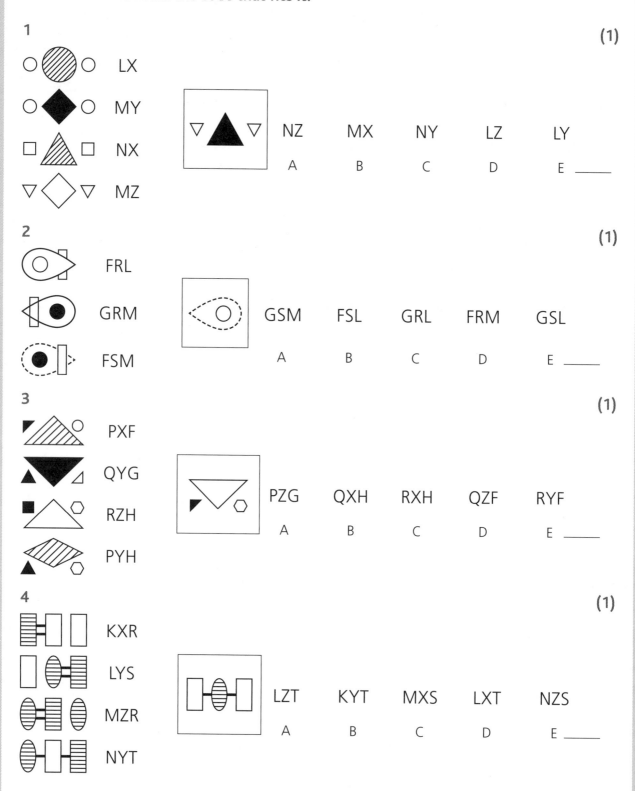

1 (1)

LX

MY

NX

MZ

	NZ	MX	NY	LZ	LY
	A	B	C	D	E _____

2 (1)

FRL

GRM

FSM

	GSM	FSL	GRL	FRM	GSL
	A	B	C	D	E _____

3 (1)

PXF

QYG

RZH

PYH

	PZG	QXH	RXH	QZF	RYF
	A	B	C	D	E _____

4 (1)

KXR

LYS

MZR

NYT

	LZT	KYT	MXS	LXT	NZS
	A	B	C	D	E _____

5 (1)

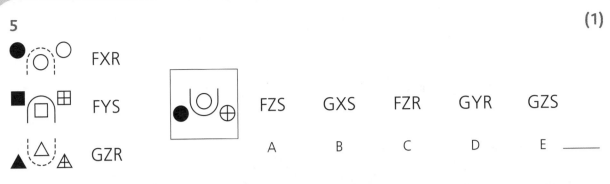

FXR

FYS

GZR

FZS   GXS   FZR   GYR   GZS

A   B   C   D   E ____

Questions 6–12: The boxes on the left or above show a pattern that is arranged in a sequence. Find the answer option that completes the sequence when inserted in the blank box.

6 (1)

7 (1)

8 (1)

9 (1)

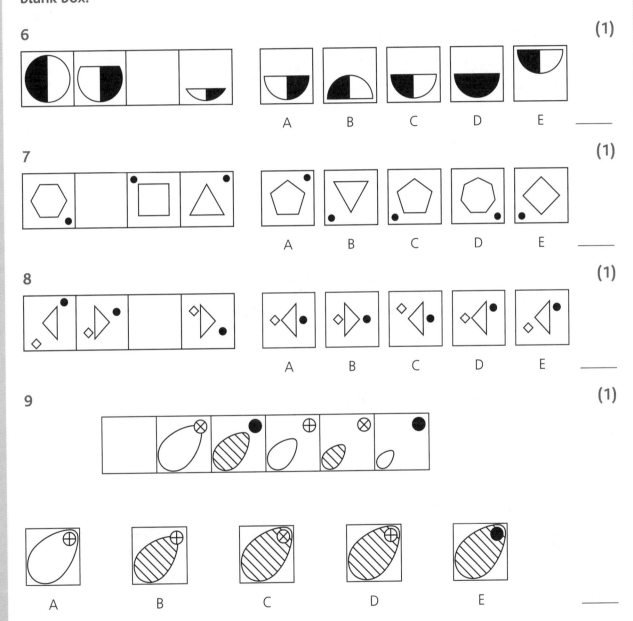

10 (1)

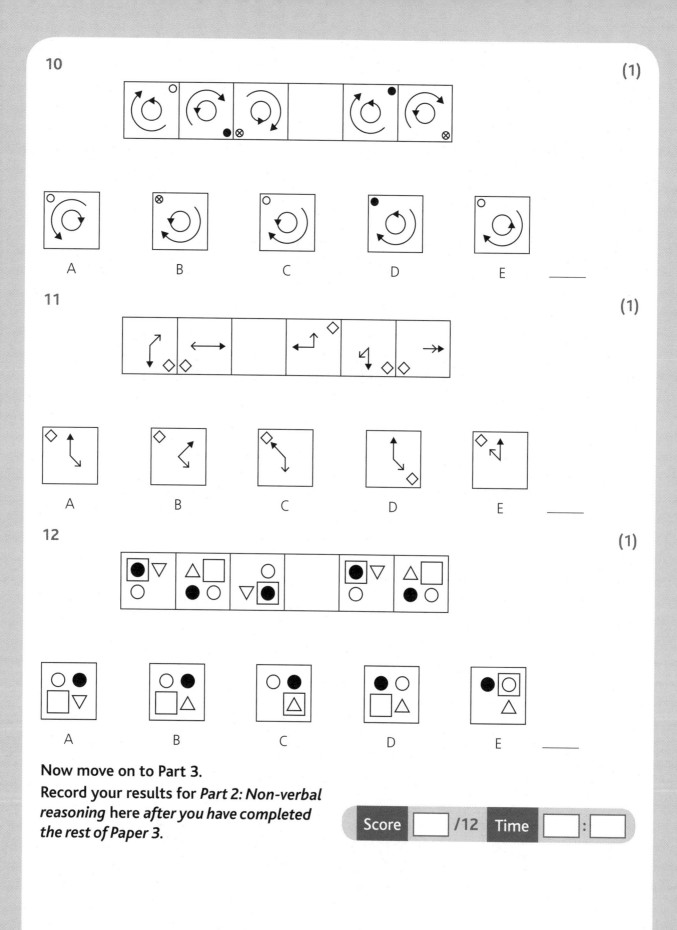

11 (1)

12 (1)

Now move on to Part 3.

Record your results for *Part 2: Non-verbal reasoning* here *after you have completed the rest of Paper 3.*

Score ☐ /12   Time ☐ : ☐

Part 3: Spatial reasoning

How to answer these questions

All your answers should be recorded on this paper. Record the letter for the answer you have chosen on the line provided. Look at the examples and then complete the training questions, which have answers at the end of this introduction. **Do not begin timing yourself until you have finished these pages.**

Only check your answers after completing all of Paper 3. The answers are in a cut-out section at the end of the book. Complete the 'results' boxes at the end of this part when you have added up your score. If you run over the time given, complete the questions and note the time you have taken.

Example question 1

Find the cube that can be made from the net shown on the left.

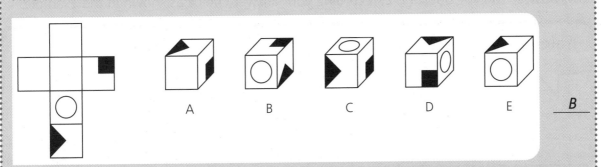

B

Answer: **B**   When the net is folded to make a cube, the triangle will point towards the face with the black square and the face with the circle will be to the left of the triangle. This makes option E impossible as the triangle points towards a blank face and the circle is to the right of the triangle. When the net is folded to make a cube, the face with the black square and the face with the triangle will be adjacent; the corner square will sit along the edge shared by both faces. This makes options A, C and D impossible.

Now answer Training question 1, recording your answer choice as shown in the example.

Training question 1

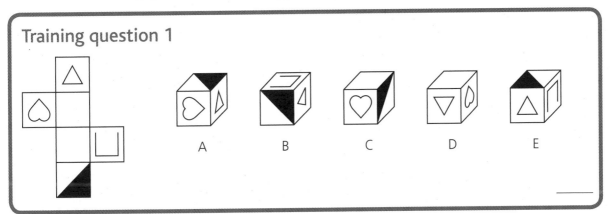

Example question 2

The picture on the left is rotated. Find the option on the right that represents this picture after it has been rotated.

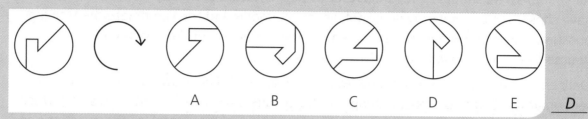

A   B   C   D   E   *D*

Answer: **D**   **rotation** – the picture is rotated 135° clockwise; **shape** – the shape of the image inside the circle does not change; **reflection** – the picture is not reflected (as in option A).

Now answer Training question 2, recording your answer choice as shown in the example.

Training question 2

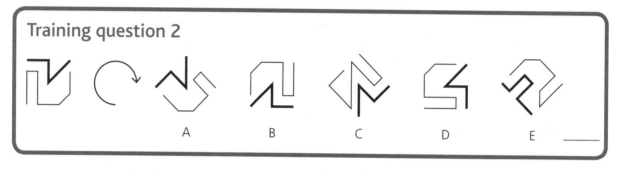

A   B   C   D   E   ___

Training question answers

1A When the net is folded to make a cube, the heart will point towards the face with the small white triangle. This makes option D impossible as, in this cube, the heart points towards a blank face. When the net is folded to make a cube, the face with the small white triangle and the face with the 'U' shape will both be adjacent to the black triangular area of the black/white face. This makes options C and E impossible since in E the 'U' is adjacent to the white triangular area and in C neither the triangle nor the 'U' borders the black triangular area. In option B, the face with the 'U' is wrong; the base of the 'U' should align with the black triangular area of the black/white face.

2B rotation – the picture is rotated 180°; **shape** – the shape of the lines does not change; **reflection** – the picture is not reflected (as in option D).

Record the letter for the answer you have chosen on the line provided.

You now have 5 minutes to complete the following 11 questions.

Questions 1–5: Find the cube that can be made from the net shown on the left.

1 (1)

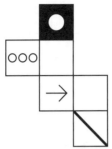

A   B   C   D   E

2 (1)

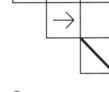

A   B   C   D   E

3 (1)

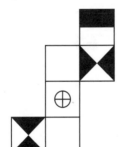

A   B   C   D   E

4 (1)

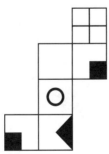

A   B   C   D   E

5 (1)

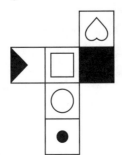

A   B   C   D   E

Questions 6–11: The picture on the left is rotated. Find the option on the right that represents this picture after it has been rotated.

6 (1)

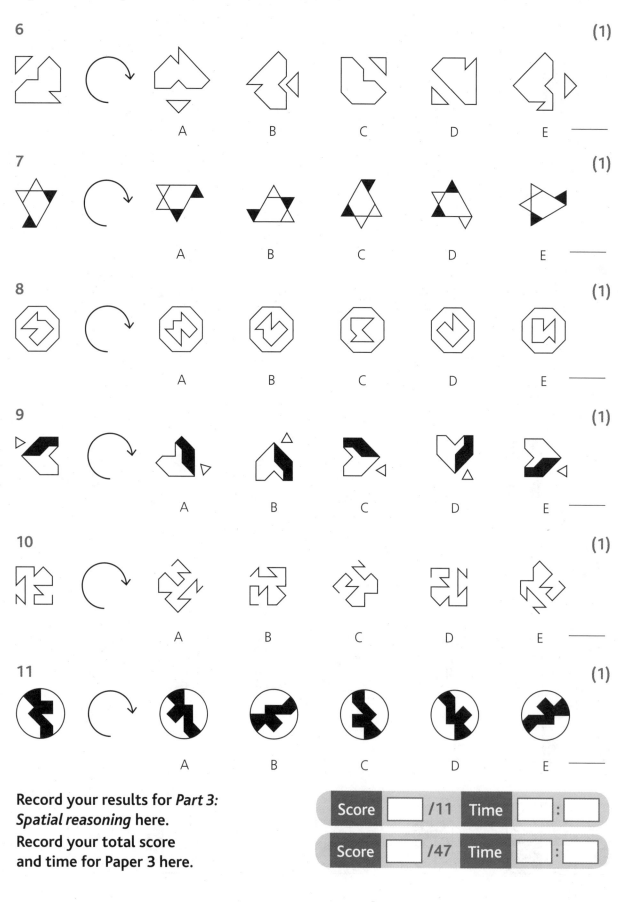

A   B   C   D   E ____

7 (1)

A   B   C   D   E ____

8 (1)

A   B   C   D   E ____

9 (1)

A   B   C   D   E ____

10 (1)

A   B   C   D   E ____

11 (1)

A   B   C   D   E ____

Record your results for *Part 3: Spatial reasoning* here.
Record your total score and time for Paper 3 here.

Score ____ /11   Time ____ : ____

Score ____ /47   Time ____ : ____

Answers

All the references in the boxes below refer to the *11+ Non-Verbal Reasoning Study and Revision Guide* (ISBN: 9781471849251) so you know exactly where to find out more about the question and your answer.

PAPER 1

Part 1: Non-verbal reasoning

1   E   **symmetry** – each picture has three straight lines, two identical small shapes and one line of symmetry. (1)

2   B   **shape/position** – each picture contains three small shapes inside a large shape: there is always a small white circle and a small white triangle; **shape/shading** – each picture contains a small black shape that is a smaller version of the large shape. (1)

3   D   **shape/number** – each picture contains a circle and a square that are connected by three straight lines to a side of the square. (1)

4   C   **shape/shading** – each large shape contains a small black triangle and a small white circle; **position/shading** – a third medium-sized shape overlaps the large outer shape once; the section of the shape that overlaps is shaded in the opposite colour to the section that is outside the large shape. (1)

5   A   **number** – each large shape is made up of six lines. *Distractors:* **shape** – the small shape is unimportant; **line style** – the line style of the large shape is unimportant. (1)

6   B   **proportion** – one-third of each picture contains a pattern or shading. (1)

7   D   **shape** – each picture contains a triangle. (1)

8   D   **position** – each picture contains an arrowhead that is fully enclosed by a straight line on each side; **direction** – the arrow points towards the middle line of this enclosure. (1)

> For more on solving and building skills to solve questions involving finding relationships, see pages 26–35.

9   C   **colour** – (a) the shading of the square alternates between black and white, (b) the shading of the arrowhead follows this repeating pattern: white, white, black, black, white, white, and so on; **translation/direction** – the square moves around the box from corner to corner in an anticlockwise direction; **rotation/direction** – the arrow rotates 90° anticlockwise and moves around the square from corner to corner in an anti-clockwise direction. (1)

10 B   **rotation/direction** – the short diagonal line attached to the outside of the square rotates 90° clockwise each time and moves from corner to corner of the square in a clockwise direction; **shading/pattern** – the pattern inside the square alternates between a cross and an 'X' shape; **shading/direction** – the quarter shading of the square with the cross pattern moves round in a clockwise direction (alternate boxes), (b) the quarter shading of the square with the 'X' pattern also moves round in a clockwise direction (alternate boxes). (1)

11 D   **shading** – the shading/pattern of the circle follows the repeating pattern: black, white, 'X' shape, and so on; **rotation** – the line plus circle rotates 90° clockwise each time; **translation/direction** – the black dot moves around the sides of the box in an anticlockwise direction, half the length of a side each time. (1)

12 B   **rotation** – the whole picture rotates 90° anticlockwise each time; **shading/position** – the shading/pattern of the right-hand section of the 'L' shape follows the repeating pattern: diagonal stripes, black, white, and so on. (1)

13 D   **rotation** – (a) the picture in the top pattern rotates 90° anticlockwise each time, (b) the picture in the bottom pattern rotates 90° anticlockwise each time; **number** – the number of dots increases by one each time (whole pattern); **number/shading** – the number of black dots increases by one each time in the top pattern and by one each time in the bottom pattern. (1)

14 E   **shape** – shapes follow the repeating pattern: circle, hexagon, square, and so on; **rotation/direction** – the striped strip rotates 90° clockwise each time around the edge of the box; **shading** – (a) the shapes on the bottom are black, (b) the shapes on the top have a cross; **shading/direction** – the shaded section in the shape with the cross pattern moves round in a clockwise direction. (1)

15 A   **shading** – the shading of the large right-angled triangle in the top row is repeated in the shape that follows it in the bottom row; **shape/size** – the shape in the bottom row is a larger version of the small shape that precedes it in the top row; **rotation/direction** – the right-angled triangle in the top row rotates 90° clockwise each time and moves from corner to corner of the box in a clockwise direction. (1)

> For more on solving and building skills to solve questions involving sequences, see pages 78–87.

Part 2: Spatial reasoning

1 **C** When the net is folded to make the cube, the face with the diagonal line will be opposite the face with the two black triangles, so option **A** is impossible. Option **E** can be discounted because a black triangle would be directly above the square with the cross, not a white triangle. In option **B**, if the square was the top face and the circle face was at the front, the face on the right would be the diagonal line, not the face with the two black triangles. In option **D**, the diagonal line is in the wrong orientation. (1)

2 **E** When the net is folded to make a cube, the face with the arrow will point towards the side of the 'U' shape, so option **B** is impossible. In option **A**, if the 'U' face and the face with the black circle were arranged as shown, the front face would contain an arrow. In option **C**, the arrow is in the wrong orientation (it would point to the left). In option **D**, if the faces with the black circle and the three white dots were arranged as shown, the bottom face would not be blank but would contain an arrow pointing to the left. (1)

3 **C** When the net is folded to make the cube, the two faces with the incomplete square shapes will be opposites, so option **B** is impossible. In option **A**, if the tall black triangle was arranged as shown with the incomplete square shape on its right, the top face would be blank and the other black triangle would be on the bottom of the cube. In option **D**, if the two faces with the black triangles were arranged as shown, the bottom face would not be blank but would contain an incomplete square. In option **E**, the tall black triangle would appear on the top face if the other faces were arranged as shown. (1)

4 **A** In option **B**, the black and white triangular sections of the face on the left would need to be switched to make this cube correct. In option **C**, the bottom face would not contain the square but would be blank. In option **D**, the triangle should not point away from the square (in the net, the triangle points towards a blank face and away from another blank face). In option **E**, the position of the shading on one of the faces is incorrect. (1)

5 **B** In option **A**, all faces are correct except for the face with the three lines, which should be horizontal. In option **C**, the face with the triangle and the face with the 'X' would need to swap places for this 3-D shape to be correct. In option **D**, all faces are correct apart from the top face as the black triangle is in the wrong position. Option **E** can be discounted because not only is the black triangle on the top face in the wrong position, but the face with the three lines would never be adjacent to a face with an arrow. (1)

6 **D** When the net is folded to make the cube, the faces with the heart, the triangle and the corner square will all be adjacent. This is the case in option **A**, but all three faces are in the wrong orientation. When the net is folded to make the cube, the arrow should point towards the heart, meaning that options **B** and **E** can be discounted. In option **C**, the face on the left side of the cube would contain the corner square not the circle. (1)

> For more on solving and building skills to solve questions involving nets and 3-D shapes, see pages 36–41.

7 **C** **rotation** – the picture has been rotated 225° clockwise; **shape/size** – the shape and size of the picture does not change. (1)

8 **E** **rotation** – the picture has been rotated 240° clockwise; **shading/position** – the position of the shading within the shape does not change. (1)

9 **D** **rotation** – the picture has been rotated 180° clockwise; **direction** – the direction of the arrow does not change; **shape** – the shape of the arrow does not change; **shading/position** – the position of the black square does not change. (1)

10 **A** **rotation** – the picture has been rotated 135° clockwise; **position** – the positions of the white dots and the shaded triangular areas of the picture do not change. (1)

11 **B** **rotation** – the picture has been rotated 90° clockwise; **shading** – the shading of the elements making up the picture does not change; **line style/angle** – the style and angle of the lines do not change. (1)

12 **B** **rotation** – the picture has been rotated 135° clockwise; **shading/position** – the position of the shading within the shape does not change; **angle/direction** – the angle/direction of the lines within the shape does not change. (1)

13 **D** **rotation** – the picture has been rotated 180° clockwise; **shading/position** – the position of the shading does not change; **shape** – the shape of the picture does not change. (1)

14 **E** **rotation** – the picture has been rotated 180° clockwise; **position** – the small squares within the large shape do not change position; **shape** – the shape of the picture does not change. (1)

> For more on solving and building skills to solve questions involving rotation, see pages 54–55.

15 **A** One 'T' shape is placed to the back left and the other is tipped to its right and placed such that the left edge sits under the extended part of the first 'T' shape. The 'L' shape is rotated 180° and sits in front to the left. (1)

16 **D** One of the single cubes slots into the gap in the 'U' shape to make a 3 × 2 cuboid. The 3 × 1 cuboid tips over onto its side and fits directly in front of the 3 × 2 cuboid. The remaining single cube fits on top of the 3 × 1 cuboid on the far left. (1)

17 E The 'S' shape forms the base and sits at the back of the stack. One of the cuboids turns and sits beneath the right side of the 'S' shape while the other cuboid sits in front and to the left. The single cube sits on top at the far right. Note that the components in option A, when assembled, would make a model that would match the shape of the target; however, there is no single cube in option A, but in option E, there is a single cube that sits at the top. (1)

18 C One of the cuboids fits on top of the long 'L' shape to make a 3 × 2 × 1 cuboid shape. The short 'L' shape sits in front of the 3 × 2 × 1 cuboid shape on the far right. The remaining cuboid slots in at the front, on the left of the short 'L' shape. (1)

19 B The 'L' shape facing away from the front is tipped backward and then slots together with the side facing the 'L' shape. The single cube sits on top of the side facing the 'L' shape. The long cuboid slots in on the right side of the model. The short cuboid is tipped sideways so that it stands upright and then sits on top of the long cuboid at the back. In option E, the gap would be beneath the 'L' shape, thereby giving nothing for the 'L' shape to rest on. This makes E an impossible construction. (1)

20 C The corner shape made up of four cubes is tipped forward, so that the three cubes are at the top and the single cube is at the bottom. The 'L' shape rotates 90° clockwise and slots together with the corner shape at the back. One of the cuboids fits into the gap at the front, with one-half fitting into the gap and the other half extending out. The other cuboid slots in at the back, behind the front cuboid. (1)

21 D The short 'L' shape is tipped forward and then rotates 90° anticlockwise. A single cube fits into the gap in the short 'L' shape to make a 2 × 2 × 1 cuboid shape. The long 'L' shape fits alongside the 2 × 2 × 1 cuboid shape on the left. The cuboid slots in next to the long 'L' shape on the left. The remaining two single cubes sit on top on the cuboid, one cube at each end. (1)

22 D One of the cuboids is tipped forward and rotated 90°; this slots in on the right side of the 'S' shape on the bottom layer of the model. Only one half of this cuboid is visible because the 'L' shape is placed at the front of the model; the top right part of the 'S' shape is also obscured by the 'L' shape. Another cuboid is tipped forward and slots in on the other side of the 'S' shape on the bottom layer of the model; only one half of this cuboid is visible because the third cuboid (tipped forward and rotated 90°) slots in above it on the left side of the 'S' shape. (1)

For more on solving and building skills to solve questions involving assembling 3-D figures, see pages 60–63.

Part 3: Non-verbal and spatial reasoning

1 B The complete picture is reflected in a vertical mirror line so the whole picture flips to the right. In option E, only one part of the picture has been reflected; in options C, D and E, the shape of the picture does not match the original. (1)

2 E The complete picture is reflected in a vertical mirror line so the whole picture flips to the right. Note that, as a result, the black shading will be in the opposite half of the circle and the small triangle will be on the left side of the large semi-circle instead of on the right. In option C, the shading appears as it would in a reflection and the small triangle appears on the correct side of the large semi-circle, but note that the small triangle has not been flipped to the right. (1)

3 D The complete picture is reflected in a vertical mirror line so the whole picture flips to the right. Note that, as a result, the black shading will be in the opposite half of the triangle. In options B and E, the shading appears as it would in a reflection, but the shape of the picture in both of these answer options does not match the original picture. (1)

4 E The complete picture is reflected in a vertical mirror line so the whole picture flips to the right. Note that, as a result, the square with the cross shape and the triangle (at the bottom of the picture) will switch places, while the square with the 'X' shape will appear at the left end of the rectangle at the top of the picture instead of on the right. Although option C shows the above elements as they would appear in a reflection, the two middle lines have switched places. (1)

5 C The complete picture is reflected in a vertical mirror line so the whole picture flips to the right. Note that, as a result, the circle and the leaf shape will switch sides and the small black rectangle and the short horizontal line at the bottom will also switch places. Although options A, B and E show the above elements as they would appear in a reflection, one or more shading patterns are incorrect. (1)

6 A The complete picture is reflected in a vertical mirror line so the whole picture flips to the right. In option B, the shading of the circles is incorrect; in C, the shading of the squares is incorrect; in D, the inner triangle is missing; and in E, the shape of the picture does not match the original. (1)

7 C The complete picture is reflected in a vertical mirror line so the whole picture flips to the right. Note that, as a result, the arrow will point in the opposite direction and the square with the 'X' shape will appear at the top right instead of the top left. Although option E shows the above elements as they would appear in a reflection, the zig-zag line at the bottom has been flipped on a horizontal plane. (1)

For more on solving and building skills to solve questions involving reflection and symmetry, see pages 50–53.

8   **B**   **shape** – the shape from the top row and the shape from the middle row in each column appear together in the same box in the bottom row; **position** – the shapes retain their original position when they appear together in the bottom row box (top shape above the middle shape); **shading** – the shapes from the top and middle rows in each column switch shading patterns when they appear together in the bottom row.   (1)

9   **E**   **shading/pattern** – the patterns inside the large shapes match in a diagonal pattern from top left to bottom right; **shape** – the large shapes match in a diagonal pattern from top right to bottom left; **direction/translation** – working in a vertical pattern from top to bottom, the small shapes move from one corner to the next in a clockwise direction.   (1)

10   **D**   **shape/size** – the large shapes are enlarged copies of the inner shapes within the four centre circles.   (1)

11   **A**   **number** – (a) the number of dots in the four triangles positioned against the outer edges of the octagon increases by one each time, moving in a clockwise direction, (b) the number of rings around the point of each triangle increases by one each time, moving in an anticlockwise direction.   (1)

12   **E**   **shading** – alternate inner points of the triangular sections of the hexagon are shaded black; **line style** – alternate outer edges of the triangular sections of the hexagon have a thick black line; **number** – the number of arrows increases by one each time, moving in a clockwise direction; **direction** – the direction of the arrows in each triangular section alternates between pointing inwards and pointing outwards.   (1)

For more on solving and building skills to solve questions involving matrices, see pages 88–97.

PAPER 2

Part 1: Spatial reasoning

1   **D**   Diagram D falls onto its right side and is then rotated 180°.   (1)
2   **A**   Diagram A is tipped backwards and then rotated 90° anticlockwise.   (1)
3   **E**   Diagram E is rotated 90° clockwise.   (1)
4   **F**   Diagram F is tipped backwards and then rotated 90° anticlockwise.   (1)
5   **B**   Diagram B is tipped onto its left side.   (1)
6   **C**   Diagram C simply falls directly backwards.   (1)

For more on solving and building skills to solve questions involving rotations with 3-D shapes, see pages 56–57 and 60–63.

7   **D**   The bottom layer of the diagram has two cubes at the front. In the middle row, there are also two cubes, one at each end of the row with two gaps in the middle; the corners of these two cubes touch the corners of the cubes on the front row. The stack of cubes on the back row touches the corner of the cube on the left in the middle row. At the top of the stack, there is another cube aligned with the back row, opposite the second cube on the front row. When looked at in plan view, this will show one square at either end of the middle row and two squares in the centre of the front and bottom rows.   (1)

8   **A**   To the far left of the diagram there are four stacked cubes; these form a line of three squares from top to bottom when viewed on the plan. There is a gap on the front row and then two cubes. At the end of the middle row, there is a stack of four cubes, at the top of which another two cubes sit on the back row. When looked at in plan view, this will show a reversed 'C' shape to the far right of the plan.   (1)

9   **B**   The bottom layer of the diagram has three cubes at the back. In the middle row, there is one cube next to the second one at the back, then a gap to another cube in the middle row. One cube sits next to this gap in the front row. At the top of the stack, at the end of the middle row, another cube is located on the far back corner. When looked at in plan view, this will show four squares on the top row, two squares on the middle row starting from the second in from the left, and then a gap to the second shaded square. On the bottom row, only the third square from the left will be shaded.   (1)

10   **C**   The bottom layer of the diagram has three cubes at the front. There is one cube at either end of the middle row and one cube on the back row, second from the left. At the top of the stack to the far right of the diagram, two cubes stick out into the front and back rows. When looked at in plan view, this will show the middle two squares unshaded and all outer squares shaded, except for the top left and the third in from the left on the back row.   (1)

11   **C**   The back row consists of a stack of cubes (three cubes tall) to the far left and then one cube extending out to the right from the top cube in this stack. The middle row has a gap on the far left and then one cube (extending out from the second cube in on the left of the back row); there is then a gap before a single-cube stack that is four cubes tall and is partly hidden by the stack on the front row. The front row has two cubes on the bottom layer, followed by a stack of three cubes with one cube extending out to its right. When looked at in plan view, this will show two squares at the top left of the plan, two squares in the middle row (second square in on the left and far right square), and all four squares shaded on the bottom row.   (1)

12 E The back row of the diagram has three cubes: two on the left on the bottom layer, a gap, and then one on the far right that is the top cube in a stack of three. In the middle row, there is a stack of four cubes on the far right, a gap, then a cube second from the left on the bottom layer and then another gap on the far left. In the front row, there are three cubes on the bottom layer and then one at the far right extending from the top of the stack on the middle row. When looked at in plan view, this will show three squares in the far right column, four squares in the bottom row, two squares in the top left of the top row and one more shaded square in the middle row, two in from the left. (1)

For more on solving and building skills to solve questions involving 2-D views of 3-D pictures, see pages 40–41.

Part 2: Non-verbal and spatial reasoning

1 E **rotation** – the whole picture rotates 180°; **line style** – the large 'L' shape and the small 'L' shape swap line styles; **direction** – the arrow flips 180°. (1)

2 D **rotation** – the whole picture rotates 90° anticlockwise; **shading/line style** – the large shape adopts the shading pattern of the small shape on the right and the small shape on the right adopts the shading pattern/line style of the large shape; **position/size** – the small shape on the left moves inside the small shape on the right, with both increasing in size. (1)

3 B **position/size** – the three small shapes at the top of the picture increase in size and move inside each other; **shading** – the central shape then takes its shading from the shape at the base of the original picture; **line style** – the middle-sized shape adopts its line style from the original far right small shape and the largest of the new shapes takes its line style from the vertical rule. (1)

4 A **size/position/shading** – the new bottom shape is taken from the centre right shape, which increases in size and adopts the shading colour/pattern of the shape on the left; the outer shape above is taken from the middle left shape, increasing in size with its shading pattern removed; the bottom shape in the original picture decreases in size, moves to the centre of the new top shape and adopts the shading colour/pattern of the original centre right shape; **reflection/position** – the top shape in the original picture reflects on the horizontal then moves inside the new top shape. (1)

5 B **line style** – the top left corner line and the bottom right corner line switch line styles; **shape/proportion** – the top left corner line (proportion of a regular shape) expands to form the whole shape; the bottom right corner line (proportion of a regular shape) expands to form the whole shape; **shading** – the small bottom right shape adopts the colour/pattern of the small bottom left shape; **position** – the bottom right shape moves inside the new whole shape formed by the expansion of the bottom right corner line; **size/position** – this new whole shape decreases in size and then moves inside the new whole shape formed by the expansion of the top left corner line. (1)

6 D **size/proportion** – the bottom large shape increases in size but then its top half is removed; **shading** – the top left circle adopts the colour/pattern of the top right triangle; **rotation** – the line rotates 135° anticlockwise; **translation/position** – the line moves to sit on top of the large shape (with a slight overlap) and the circle moves to the centre of the line. (1)

7 C **shading** – the small rectangle on the left and the small rectangle on the right switch shading patterns; **reflection** – the triangular shape flips on a horizontal plane; **size** – the triangular shape increases in size, covering most of the original large shape. (1)

For more on solving and building skills to solve questions involving applying changes to shapes, see pages 26–35, 46–49 (position and direction), 50–53 (reflection and symmetry) and 54–55 (rotations).

8 D When the net is folded to make the cube, the face with the cross will be on the opposite face to the face with the star. Imagine folding together the face with the cross and the face with the circle so that the circle becomes the top face, and then folding down the face with the clover shape. When you fold down the face with the star, it will be opposite the face with the cross. (1)

9 E When the net is folded to make the cube, the face with the two white triangles will be opposite the face with the four black and white triangles. Imagine folding together the face with the two white triangles and the face with the octagon so that the face with the octagon becomes the top face, then folding down the face with the 'H' shape. When you fold down the face with the four black and white triangles, it will be opposite the face with the two white triangles. (1)

10 B When the net is folded to make the cube, the face with the star will be opposite the face with the diagonal striped pattern. Imagine folding down the face with the star, the face with the cross and the face with the circle so that the face with the 'flying saucer' shape becomes the top face. When you fold round the face with the diagonal striped pattern, it will be opposite the face with the star. (1)

11 A   When the net is folded to make the cube, the face with the two white circles will be opposite the face with the arched shape. Imagine folding together the face with the arched shape and the face with the hatched shading pattern so that the face with the hatched shading pattern becomes the top face. Then imagine folding down the face with the four small squares. When you fold down the face with the two white circles, it will be opposite the face with the arched shape. (1)

12 D   When the net is folded to make the cube, the face with the circle will be opposite the face with the cross. Imagine folding together the face with the circle and the face with the small circle inside a triangle so that the face with the small circle inside a triangle becomes the top face. Then imagine folding down the face with the heart shape. When you fold down the face with the cross, it will be opposite the face with the circle. (1)

13 E   When the net is folded to make the cube, the face with the circle and four corner triangles will be opposite the face with the hexagon. Imagine folding together the face with the circle and four corner triangles and the face with the wavy shape so that the face with the circle and four corner triangles becomes the top face. Then imagine folding down the face with the octagon. This leaves the face with the hexagon free to fold underneath the cube so that it is opposite the face with the circle and four corner triangles. (1)

For more on solving and building skills to solve questions involving nets and 3-D shapes, see pages 36–41.

Part 3: Non-verbal reasoning

1 C   **shape** – in four of the pictures the central small shape is a regular polygon. Only the circle in C contains an irregular shape.
Distractor: **line style** – the line style of the outer circles is unimportant. (1)

2 E   **shape** – only E does not contain a triangle.
Distractors: **shading** – (a) the shading of the overlapping shape is unimportant, (b) the shading of the small circle is unimportant; **position** – the position of the circle is unimportant. (1)

3 B   **position** – in four of the pictures the striped triangle is positioned outside the lines that make the large open triangle; only the triangle in B is positioned inside the lines.
Distractor: **shading/pattern** – (a) either square can have a cross pattern or an 'X' shape, (b) the shading pattern of the triangle is unimportant. (1)

4 D   **shading** – in four of the pictures the overlapping shape is shaded outside the large shape; only D contains an overlapping shape where the shading is inside the large shape.
Distractors: **position** – the position of the white shape is unimportant; **direction** – the direction of the arrow is unimportant; **size/length** – the length of the arrows is unimportant. (1)

5 E   **rotation** – four of the pictures are identical and would fit exactly on top of each other if they were rotated round to the same position; E is the only picture that has been flipped. (1)

6 C   **shape** – four of the pictures contain a large curved shape; the large shape in C is the only shape that has a straight edge.
Distractor: **shape** – the smaller shapes are unimportant. (1)

For more on solving and building skills to solve 'most unlike' questions, see pages 26–35.

7 A   **direction** – two of the eye shapes 'look' up while the third eye shape 'looks' in a different direction from the other two. (1)

8 B   **line style** – in each picture there are four lines: a wavy line, a zig-zag line, a straight line and a 'stitched' line. These lines can be in any order. One of these lines is bold. (1)

9 D   **number** – (a) the large shape in each picture is made up of five lines, (b) the large shape contains two small shapes; **shape** – one small shape is curved while the other is straight-lined; **shading** – one of the small shapes is black and the other is white. (1)

10 A   **symmetry** – each picture has one line of symmetry.
Distractor: **number** – the number of shapes is unimportant. (1)

11 C   **shape/size** – each picture contains a large shape and a smaller version of the large shape; **line style** – the outlines of the two shapes are different.
Distractor: **position** – the overlap is unimportant. (1)

12 E   **number** – each picture contains a large shape made up of eight lines.
Distractor: **position** – the position of the small shape is unimportant. (1)

13 C   **shape/position** – each picture contains a triangle in the middle of a line and two circles at either end; **shading/direction** – the triangle points towards a black circle; the circle at the other end of the line is white. (1)

14 E   **position** – (a) the black corner square is never positioned where the two large squares join, (b) the white corner square must touch the join of the two large squares. Therefore, the two shaded corner squares must not touch. (1)

15 D **direction** – each picture contains one large arrow and one smaller arrow pointing in the same direction.
Distractors: **shape** – the shape of the arrows is unimportant; **shading** – the shading of the arrows is unimportant. (1)

16 A **shape/number** – each picture contains two triangles and a six-sided shape.
Distractors: **shading** – the shading of the shapes is unimportant; **position** – the position of the shapes is unimportant; **line style** – the line style of the shapes is unimportant. (1)

For more on solving and building skills to solve questions involving finding relationships, see pages 26–35 and 50–53 (reflection and symmetry).

17 C **shape** – the first letter represents both shapes in each pair: **R** is a pair of identical shapes, while **S** is a pair of non-identical shapes; **shading** – (a) the second letter represents the shading of the shape on the left in each pair: **X** is diagonal stripes, **Y** is black and **Z** is white, (b) the third letter represents the shading of the shape on the right in each pair: **F** is horizontal stripes, while **G** is white. (1)

18 B **direction** – the first letter represents the direction of the arrow: **W** is right, **X** is up and **Y** is left; **shading** – the second letter represents the shading of the arrowhead: **L** is black and **M** is white. (1)

19 A **shape** – (a) the first letter represents the small shape: **P** is a circle and **Q** is a triangle, (b) the third letter represents the large shape: **L** is a heart, **M** is an irregular pentagon and **N** is an arched shape; **shading** – the second letter represents the shading of the small shape: **X** is black, while **Y** is diagonal stripes. (1)

20 D **angle/position** – the first letter represents the position of the right angle in the right-angled triangle: **G** is top left, while **H** is top right; **number** – the second letter represents the number of vertical lines inside the right-angled triangle: **S** is three and **T** is one.
Distractor: **shape** – the small shape is unimportant. (1)

21 D **shape** – the first letter represents the large outer shape: **P** is a quadrilateral with two right angles, while **Q** is a rectangle and **R** is a trapezium; **line style** – the second letter represents the line style used in the large shape: **X** is a thick line, while **Y** is a thin line and **Z** is a dashed line.
Distractor: **number** – the number of squares inside the large shape is unimportant. (1)

22 C **number** – the first letter represents the number of triangles in each picture: **L** is three, while **M** is two and **N** is one; **line style** – the second letter represents the line directly above the triangles: **R** is a wavy line, while **S** is a thick solid line and **T** is a dashed line.
Distractors: **shading** – the shading of the triangles is unimportant; **line style** – the line style of the top line is unimportant. (1)

23 E **shape** – the first letter represents the small white shape on the left: **R** is a circle, **S** is a triangle and **T** is a shield shape; **line style** – the second letter represents the line style of the inner shape on the right: **F** is a bold line, while **G** is a dashed line and **H** is a thin line; **position** – the third letter represents the position of the open sides of the large shapes: **L** is mixed (the left shape has an open side at the top and the shape on the right has an open side at the bottom), while **M** is top (both shapes are open at the top) and **N** is bottom (both shapes are open at the bottom). (1)

24 A **shape** – the first letter represents the inner shape: **X** is an ellipse, while **Y** is a rectangle and **Z** is an irregular pentagon; **shading** – the second letter represents the colour/pattern of the inner shape: **P** is vertical stripes, while **Q** is black, **R** is white and **S** is dotted; **line style** – the third letter represents the line style of the large outer shape: **F** is a thin line, while **G** is a thick solid line and **H** is a dashed line. (1)

For more on solving and building skills to solve questions involving connections with codes, see pages 68–77.

PAPER 3

Part 1: Non-verbal and spatial reasoning

1 E There are a number of options as to how the long cuboid, short cuboid and small cube are stacked at the back of the diagram. One way is for the short cuboid to be tipped back so that it sits on its long edge and it then fits on top of the long cuboid, on the left. The single cube sits on top of the short cuboid, on the left. The 'L' shape rotates 90° anticlockwise and slots in at the front of the model, on the left. (1)

2 A One of the corner shapes made up of four cubes flips over and joins with the other corner shape to make a 2 × 2 × 2 cube. One of the single cubes sits at the front of this large cube on the left, one on the right of the large cube at the back and the remaining single cube sits on top of the large cube at the back, on the right. (1)

3 B The 'T' shape is tipped to the left so that it stands on its side with the base of the 'T' sticking out to the right. One of the 'L' shapes wraps around the base of the 'T' shape. The cuboid is placed to the right of the combined 'T' and 'L' shapes. The remaining 'L' shape is turned sideways and sits to the far right end of the model. (1)

4 C The three single cubes are stacked on top of each other and placed at the right end of the 'L' shape. The first cuboid slots in front of the stack of single cubes. The second cuboid is tipped forward, rotated 90° and fits in front of the 'L' shape. The third cuboid is tipped forward and slots in between the other two cuboids. (1)

5   E   The first 'T' shape is flipped 180° and the 'L' shape is rotated 90°: these two shapes slot together, with the 'T' shape on the right and the 'L' shape on the left. The second 'T' shape is tipped backwards so that it stands on its side with the base of the 'T' sticking out to the left; it is then rotated 90° clockwise. This second 'T' shape slots in between the 'L' shape and the first 'T' shape, with the protruding 'base' of the 'T' fitting into the central gap on the second layer of the model.   (1)

6   A   The cuboid is rotated 90° clockwise and slots through the left gap of the corner shape so that only one half of the cuboid is visible. The first single cube fits directly behind the cuboid on the bottom layer of the model (note, this cube is not visible in the diagram of the model). The second single cube is stacked on top of the first single cube (at the back, on the left side of the model). The third single cube slots into the right gap of the corner shape, next to the other single cube on the bottom layer of the model.   (1)

For more on solving and building skills to solve questions involving assembling 3-D figures, see pages 60–63.

7   D   **number/size** – each picture contains one triangle and three straight lines of differing lengths; **position** – the triangle is positioned at the end of the shortest line.   (1)

8   C   **number** – (a) the large shape in each picture is made up of six lines, (b) there are two small shapes in each picture.   (1)

9   B   **shape** – the inner and outer shape are the same shape; **line style** – the outer shape has a solid outline, while the inner shape has a dashed outline; **number** – each picture contains just one small shape.   (1)

10   D   **shading/position** – (a) in each picture the large shape contains one small black shape and one small white shape, (b) outside the large shape is a small white shape.
Distractors: **line style** – the line style used for the outline of the large shape is unimportant; **shape** – both the large and small shapes can be any shape.   (1)

11   E   **number** – the large shape in each picture is a quadrilateral; **shading** – each quadrilateral contains one white shaded corner and one black shaded corner; **direction/position** – (a) the arrow is positioned in one of the corners of the quadrilateral and points towards the white shaded corner, (b) the black shaded corner is positioned anticlockwise to the corner with the arrow.   (1)

12   A   **position/rotation** – if each picture was rotated round to the same position, they would all be identical: each shape and shading pattern would be in exactly the same place.   (1)

13   E   **number** – each picture contains one large white shape and two small black shapes; **position** – one of the small shapes is connected to one or more sides of the large shape, while the other small shape floats freely inside or outside of the large shape.   (1)

14   D   **symmetry** – each picture contains one line of symmetry; **shape** – (a) there is a circle inside each large shape, (b) each large shape has one or more straight edges.   (1)

For more on solving and building skills to solve questions involving finding relationships, see pages 26–35.

15   C   **reflection** – the whole picture is reflected in a vertical mirror line; **rotation** – the top small shape rotates 180°; **shading/rotation** – the shading pattern of the long shape rotates 90° and then transfers to the bottom small shape.   (1)

16   E   **translation** – the small shape or shapes at the top of the picture move inside the large shape; **rotation** – (a) the whole picture then rotates 90° clockwise, (b) the small shape or shapes rotate 90° anticlockwise; **shading** – the small shape or shapes are shaded in the opposite colour in the second picture.   (1)

17   B   **size/number** – the small shape expands to become the large shape in the second picture and the large shape decreases in size but multiplies in number; the number of new shapes created matches the number of sides in the original large shape; **shading** – the large shape and the small shape switch shading colours/patterns.   (1)

18   C   **reflection** – the top large shape is reflected in a horizontal mirror line; the two small shapes inside the large shape are not reflected; **translation** – the bottom shape moves up to the top and the two small shapes move either side of the new top shape; **shading** – any shading in the bottom shape in the first picture transfers to the two small shapes in the second picture.   (1)

19   B   **rotation** – (a) the whole picture rotates 90° clockwise, (b) the two small shapes either side of the central section in the large shape flip on a horizontal plane; **translation** – the dots in the central section of the large shape in the first picture move to the left-hand section of the large shape in the second picture.   (1)

For more on solving and building skills to solve questions involving applying changes to shapes, see pages 26–35, 46–49 (position and direction), 50–53 (reflection and symmetry) and 54–55 (rotations).

20   E   **shape** – shapes match in a horizontal pattern; **rotation** – shapes rotate 135° clockwise from box to box, working from left to right; **line style** – lines (boldness) match in a vertical pattern.   (1)

21   C   **shape** – shapes match in a diagonal pattern, working from top left to bottom right; **shading** – colours/patterns match in a diagonal pattern, working from top right to bottom left; **line style** – the line styles of the squares match in a horizontal pattern, working from left to right.   (1)

CEM 11+ Non-Verbal Reasoning & Spatial Reasoning Practice Papers published by Galore Park

22   B   **number/size** – working from top to bottom, the four shaded concentric shapes in the top row become four separate shapes of equal size in the bottom row; **position** – (a) the largest concentric shape in the top row moves to the bottom right corner of the box in the bottom row, (b) the second largest concentric shape moves to the top left corner of the box in the bottom row, (c) the third largest concentric shape moves to the bottom left corner of the box in the bottom row, (d) the smallest concentric shape moves to the top right corner of the box in the bottom row.   (1)

23   A   **shape/position** – the top shape and the middle shape in each column switch positions when they appear together in the same box in the bottom row; **shading** – the top shape and the middle shape swap shading/pattern/colour.   (1)

24   D   **number** – starting in the top left hexagon and working round in a clockwise direction, the number of triangles decreases by one each time; **shading/number** – working in a clockwise direction starting with the top left hexagon and finishing in the centre, the number of circles increases by one each time; the new circle is the opposite colour to the previous circle added to the pattern.

Distractor: **position** – the arrangement of the circles and triangles within each hexagon is unimportant.   (1)

> For more on solving and building skills to solve questions involving matrices, see pages 88–97.

Part 2: Non-verbal reasoning

1   C   **shape** – the first letter represents the large shape: **L** is a circle, while **M** is a diamond and **N** is a triangle; **shading** – the second letter represents the colour/pattern of the large shape: **X** is diagonal stripes, **Y** is black and **Z** is white.

Distractor: **shape** – the small white shapes are unimportant.   (1)

2   E   **direction** – the first letter represents the direction in which the large shape is pointing: **F** is right and **G** is left; **line style** – the second letter represents the style of line used for the large shape: **R** is a thin solid line, while **S** is a dashed line; **shading** – the third letter represents the colour/pattern of the circle: **L** is white and **M** is black.

Distractor: **shape/position** – the rectangle shape and position is unimportant.   (1)

3   C   **shading** – the first letter represents the colour/pattern of the large shape: **P** is diagonal stripes, **Q** is black and **R** is white; **shape** – (a) the second letter represents the small black shape on the left: **X** is a right-angled triangle, while **Y** is an equilateral triangle and **Z** is a square; (b) the third letter represents the small white shape on the right: **F** is a circle, while **G** is a right-angled triangle and **H** is a hexagon.   (1)

4   D   **position** – (a) the first letter represents the sequence of the shapes: **K** is rectangle, rectangle, rectangle, while **L** is rectangle, ellipse, rectangle, **M** is ellipse, rectangle, ellipse and **N** is ellipse, rectangle, rectangle, (b) the third letter represents the position of the two connecting lines: **R** is two lines between the first two shapes, while **S** is two lines between the middle and right-hand shape and **T** is one line between each shape; **number/shading** – the second letter represents the total number of shapes that are shaded: **X** is one, **Y** is two and **Z** is three.   (1)

5   B   **direction** – the first letter represents the direction of the 'U' shape: **F** is upside down, while **G** is the right way up; **shape** – the second letter represents the three small shapes: **X** is circles, **Y** is squares and **Z** is triangles; **line style** – the third letter represents the style of line used for the 'U' shape: **R** is a dashed line and **S** is a solid line.   (1)

> For more on solving and building skills to solve questions involving connections with codes, see pages 68–77.

6   C   **proportion/position** – a proportion of the circle is removed from the top of the shape each time; **shading** – the shading of the circle switches sides each time.   (1)

7   C   **direction/position** – the black dot moves around the box from one corner to the next, in a clockwise direction; **number** – the number of sides in the large shape decreases by one each time.   (1)

8   A   **direction/position** – (a) the small white diamond shape moves up the box in a vertical direction from the bottom to the top in five equal steps, (b) the small black circle moves down the box in a vertical direction from the top to the bottom in five equal steps, (c) the large triangle alternates between pointing to the left and pointing to the right.   (1)

9   D   **size** – the egg shape decreases in size each time; **shading** – (a) the shading colour/pattern of the egg shape alternates between stripes and white shading, (b) the shading colour/pattern of the circle follows this repeating pattern: cross, 'X' shape, black, cross, and so on.   (1)

10   E   **direction/position** – the small circle moves around the box from one corner to the next in a clockwise direction; **shading** – the colour/pattern of the small circle follows this repeating pattern: white, black, 'X' shape, and so on; **rotation** – (a) the curved arrow rotates 90° clockwise each time, (b) the central circle rotates 90° anticlockwise each time.   (1)

11 A **direction/position** – the small diamond moves around the box from one corner to the next in a clockwise direction; **rotation** – (a) the arrow with the solid arrowhead rotates 90° anticlockwise each time, (b) the arrow with the open arrowhead rotates 135° anticlockwise each time. (1)

12 B **direction/position** – (a) the black circle moves around the box one space at a time in an anticlockwise direction, (b) the white square moves around the box one space at a time in a clockwise direction, (c) the triangle moves around the box one space at a time in an anticlockwise direction, (d) the white circle moves around the box one space at a time in an anticlockwise direction; **rotation** – the triangle flips 180° each time. (1)

For more on solving and building skills to solve questions involving sequences, see pages 78–87.

Part 3: Spatial reasoning

1 A When the net is folded to make a cube, the face with the arrow will point towards a blank face. This makes option **C** impossible as the arrow is pointing towards the face with the diagonal line. In option **B**, if the front face and the top face were arranged as shown, the three circles on the left face would appear in a horizontal line, not a vertical line. In option **D**, if the top face was the white circle on a black background, the three circles on the right face would be arranged in a horizontal line and the diagonal line on the front face would go from top right to bottom left. In cube **E**, if the faces were arranged as shown, the arrow and the three circles would display in the correct orientation but the diagonal line on the top face would go from top right to bottom left. (1)

2 B When the net is folded to make a cube, the face with the black circle will be opposite the blank face. This makes option D impossible. When the net is folded to make a cube, the two faces containing a large black triangle will be opposite each other. This makes option A impossible. In option C, the black triangle is pointing towards the face with the black circle, which is not possible because, in the net, a black triangle either points towards the blank face or the 'U' shape. In option E, if the faces of the cube were arranged as shown, the 'U' shape would be the other way up. (1)

3 D When the net is folded to make a cube, the face with the cross in the circle will be opposite the face with the black and white rectangles. This makes options A, B and E impossible. When the net is folded to make a cube, the two faces with black and white triangles will be opposite each other. This makes option C impossible. (1)

4 C When the net is folded to make a cube, the two faces with a black square will be opposite each other. This makes option A impossible. When the net is folded to make a cube, the black triangle will point towards one of the faces with a black square; the black square will appear to be in the top left corner of its face when looked at in relation to the face with the black triangle. This makes option B impossible. In option D, if the face with the cross was the right face (as shown in the diagram), the triangle on the front face would point down towards the face with the black square, and the black square on the bottom face would be in the bottom right corner (not the bottom left). In option E, if the faces of the cube were arranged as shown, the black square would be in the top right corner of its face, not the top left corner. (1)

5 E When the net is folded to make a cube, the heart shape will point towards the face with the black triangle. This makes options **B** and **C** impossible. When the net is folded to make a cube, the face with the heart and the face with the circle will be opposite each other. This makes option A impossible. In option D, if the circle was the front face, the black face would be on the bottom and the square face would be on the right. (1)

For more on solving and building skills to solve questions involving nets and 3-D shapes, see pages 36–41.

6 A **rotation** – the picture is rotated 225° clockwise. (1)
7 C **rotation** – the picture is rotated 180° clockwise. (1)
8 E **rotation** – the picture is rotated 135° clockwise. (1)
9 E **rotation** – the picture is rotated 180° clockwise. (1)
10 D **rotation** – the picture is rotated 180° clockwise. (1)
11 C **rotation** – the picture is rotated 180° clockwise. (1)

For more on solving and building skills to solve questions involving rotation, see pages 54–57.